Technology Developments in Computer Intelligence and their Applications in the Era of Industry 5.0

Smart technologies such as artificial intelligence, machine learning, and computer intelligence are shaping the future of Industry 5.0. This book delves into emerging technologies that integrate human-centric approaches with AI-driven automation, addressing key innovations in deep learning, augmented reality and IoT. Through real-world applications across industries such as healthcare, robotics, and automation, this comprehensive guide provides insights into how intelligent systems are revolutionizing efficiency, decision-making, and human-machine collaboration in the digital era.

This book:

- Serves as a critical resource for understanding and navigating the evolving landscape of intelligent technologies in Industry 5.0
- Provides necessary information on technological developments in the computer intelligence and their role in industrial revolution 5.0.
- Contains information on digital twin technology and its application in autonomous vehicles, augmented reality, advances in health care and drone technology to name a few.

It is primarily written for undergraduate, post-graduate students, as well as researchers working in the area of smart technologies which are an integral part of industrial revolution 5.0.

Smart Technologies for Engineers and Scientists

Series Editor: Mangey Ram

Applications of Mathematical Modeling, Machine Learning, and Intelligent Computing for Industrial Development
Madhu Jain, Dinesh K. Sharma, Rakhee Kulshrestha, and H.S. Hota

Technology Developments in Computer Intelligence and their Applications in the Era of Industry 5.0
Mukul Kumar Gupta, Abhinav Sharma, Vinay Rishiwal, and Meera C S

Technology Developments in Computer Intelligence and their Applications in the Era of Industry 5.0

Edited by
Mukul Kumar Gupta
Abhinav Sharma
Vinay Rishiwal
Meera C S

CRC Press
Taylor & Francis Group
Boca Raton London New York

CRC Press is an imprint of the
Taylor & Francis Group, an **informa** business

Designed cover image: Shutterstock

First edition published 2026
by CRC Press
2385 NW Executive Center Drive, Suite 320, Boca Raton FL 33431

and by CRC Press
4 Park Square, Milton Park, Abingdon, Oxon, OX14 4RN

CRC Press is an imprint of Taylor & Francis Group, LLC

ISBN: 978-1-032-94108-0 (hbk)
ISBN: 978-1-032-94109-7 (pbk)
ISBN: 978-1-003-56899-5 (ebk)

DOI: 10.1201/9781003568995

Typeset in Sabon
by SPi Technologies India Pvt Ltd (Straive)

Contents

Notes on the editors

Mukul Kumar Gupta completed his B. Tech. in Electronics and Instrumentation Engineering from UPTU Lucknow. He did his M. Tech. in control system from VJTI Mumbai and doctoral degree from the department of Electrical Engineering (Robotics and Control) of University of Petroleum and Energy Studies (UPES) Dehradun. Currently he is working as Associate Professor in the Department of Electronics and Instrumentation Engineering, Faculty of Engineering and Technology, MJP Rohilkhand University, Bareilly, Uttar Pradesh, India. Prior to that he has worked at DMCE Mumbai, DIT Dehradun and UPES Dehradun. He has 18 years of experience in academics. His major research interests include nonlinear control, Robotics and Automation and Data driven control. Particularly, he is working on various applications of robotic systems in different areas. He has published more than 50 research papers in various journals and conferences of international repute. He has supervised 5 Ph. D. scholar and 18 M. Tech. scholar in the area of robotics and control. Dr. Mukul is also a certified automation engineer from BOSCH Rexroth and is closely associated with several international journals as a reviewer and editorial board member.

Abhinav Sharma received his B. Tech. from H. N. B. Garhwal University, Srinagar, India, in 2009 and the M. Tech. and Ph. D. from Govind Ballabh Pant University of Agriculture and Technology, Pantnagar, India, in 2011 and 2016. He is working as an Associate Professor in the Department of Electrical and Electronics Engineering in University of Petroleum and Energy Studies, Dehradun. He has taught subjects such as data communication networks, microprocessor and embedded system, artificial intelligence and machine learning, communication systems. He has publications in many national and international journals and conferences. His field of interest includes adaptive array signal processing, artificial intelligence, and machine learning, optimization techniques and smart antennas.

Vinay Rishiwal is working as a professor in the department of computer science and information technology, faculty of engineering and technology, MJP Rohilkhand University, Bareilly, Uttar Pradesh, India. He obtained

B. Tech. degree in Computer Science and Engineering in the year 2000 from MJP Rohilkhand University (SRMSCET), India and received his PhD in computer science and engineering from Gautam Buddha Technical University, Lucknow, India, in the year 2011. He has 23 years of experience in academics. He is a senior member of IEEE, ACM and worked as convener, student activities committee, IEEE Uttar Pradesh Section, India. He has published more than 90 research papers in various journals and conferences of international repute. He also has 20 patents to his credit. He is a General/Conference chair of four international conferences namely ICACCA, IoT-SIU, MARC 2020 and ICAREMIT. His current research interests include wireless sensor networks, IoT, cloud computing, social networks and blockchain technology.

Meera C S is a scientist at the advanced remanufacturing and technology centre (ARTC), A*STAR Singapore. She served as the lead principal investigator for the remote-control excavator program, aiming to enhance safety and productivity in excavator operations in urban construction. With a strong background in Robotics Engineering, she has since been deeply involved in cutting-edge research and projects in construction robotics, haptics, industrial automation, artificial intelligence and deep neural networks. In her previous position as assistant manager (Computer Vision) at L&T smart world and communication, she was involved in tackling real-time challenges related to smart city and safe city applications. She has published in renowned journals and conferences in robotic systems.

Contributors

Abhinav Sharma
University of Petroleum & Energy Studies
Dehradun, India

Ajay Singh
Uttaranchal University
Dehradun, India

Akanksha Singh
Amity University Greater Noida
Greater Noida, India

Amit Kukker
SRM Institute of Science and Technology, Delhi - NCR Campus
Ghaziabad, India

B. Iftekhar Hussain
Bapatla Engineering College
Bapatla, India

B. Khaleelu Rehman
Chaitanya Bharathi Institute of Technology (A)
Hyderabad, India

Bhargav Prajwal Pathri
Woxsen University
Hyderabad, India

Brijmohan
SRMS College of Engineering Technology & Research
Bareilly, India

Chaitanya Krishna Kasaraneni
Egen Solutions
Naperville, Illinois, USA

Devendra Rawat
Amity University Madhya Pradesh
Gwalior, India

Gaurav Pandey
Neta Ji Subash University of Technology,
New Delhi, India

Harsh Vardhan Saxena
SRMS College of Engineering Technology & Research
Bareilly, India

Kalyan S Kasturi
Koneru Lakshmaiah Education Foundation
Bowrampet, Hyderabad, India

Kapil Joshi
Uttaranchal University
Dehradun, India

Kawerinder Singh Sidhu
Uttaranchal University
Dehradun, India

Konda Hari Krishna
Mohan Babu University
Tirupati, India

Krishna Vamshi Ganduri
Woxsen University
Hyderabad, India

Lipsa Das
Amity University Greater Noida,
Greater Noida, India

Manvi Mishra
SRMS College of Engineering Technology & Research
Bareilly, India

Meera C S
Advanced Remanufacturing and Technology Centre
A*STAR, Singapore

M. Nagabhushana Rao
Marri Laxman Reddy Institute of Technology and Management
Hyderabad, India

Manjeet Singh
Amity University Greater Noida
Greater Noida, India

Mohammed Abdul Nasar
University of Technology & Applied Sciences, Nizwa,
Sultanate of Oman, Oman

Mukul Kumar Gupta
Mahatma Jyotisar Phule Rohilkhand University
Bareilly, India

Nadia Mahmood Hussien
Mustansir University
Baghdad, Iraq

Nandini Prasad K. S.
Dayananda Sagar Academy of Technology and Management
Bangalore, India

Nazeerunnisa Qureshey
Muffakham Jah College of Engineering and Technology
Hyderabad, India

Nisha S. L.
Ramaiah Institute of Technology
Bangalore, India

P. Anuradha
Chaitanya Bharathi Institute of Technology (A)
Hyderabad, India

Pinisetti Swami Sairam
Woxsen University
Hyderabad, India

Priyanka Singh
SRM University
Amaravati, India

Reena Pant
Mahatma Jyotisar Phule Rohilkhand University
Bareilly, India

Sukhdeep Kaur
Uttaranchal University
Dehradun, India

Vendra Durga Ratna Kumar
SRM University
Amaravati, India

Vinay Rishiwal
MJP Rohilkhand University
Bareilly, India

Yarlagadda Anuradha
Gayathri Vidya Parishad College of Engineering
Visakhapatnam, India

Preface

Smart technologies are artificial intelligence (AI) driven systems that mimics or reproduces the behavior of the human brain. These machines process the information and optimize the performance of the system over a period of time. Nowadays AI, machine learning and robotics are an integral part of our day-to-day life and is an active area of research across every domain. Some major applications are in the next-generation healthcare system, medical imaging, automation and control, automobile, education, advanced communication, internet of things (IoT), Défense system, banking, precision agriculture, manufacturing and business development. Fifth industry revolution focuses on integration of humans with AI powered robots for creating sustainable industrial environment. The field of robotics, machine learning and AI is rapidly developing, and now a diverse set of literature is available in this area. However, with the current development in computational technologies and hardware, the available information is in such a decentralized manner, especially in the healthcare sector that as an individual a reader has to put in a lot of effort to understand these technologies for various applications. The aim of this book is to give readers information about smart technologies and its applications in the current era of industrial revolution 5.0 at one centralized place. The organization of the book is as follows:

Chapter 1 presents various fine-tuned techniques with adjusting learning rates applied on the ResNet50 pre-trained model to perform accurate prediction and classification of tumors in brain by using MRI images. The proposed modified ResNet50 model estimate maximum accuracy compared to further advance models.

Chapter 2 outlines reinforcement learning core concepts, contrasting it with traditional supervised learning. The key components of reinforcement learning such as agents, actions, rewards, and policies, are introduced, outlining how agents iteratively observe, act, and learn to improve their strategies over time.

Chapter 3 reviews and provides an in-depth analysis of design and development of metamaterials inspired PRMAs for antenna designers. The chapter highlights different techniques by which the radiation pattern of a microstrip antenna can be controlled in the desired manner by using different types of metamaterials.

Chapter 4 provides a birds-eye view of the problems associated with various computational intelligence healthcare applications from a management standpoint. The chapter demonstrates the effectiveness of Computational Intelligence in increasing the accuracy of healthcare tasks.

Chapter 5 presents the reinforcement learning mathematical methods, and practical applications, highlighting how it has the potential to transform decision-making processes in the variety of fields. The chapter investigates key reinforcement learning algorithms such as Q-learning, Temporal difference learning, SARAS, and deep reinforcement learning with their practical applications in areas like robotics, energy, environment, finance, healthcare, and autonomous vehicles.

Chapter 6 innovatively utilize a Field-Programmable Gate Array (FPGA) to control the robotic arm's movements. The chapter focuses on the FPGA-based robotic arm to pick the objects and place it in the specified locations.

Chapter 7 examines the synergy of drones with smart technologies such as artificial intelligence, machine learning, and advanced sensors that will advance the application of drones, to go from automatic delivery mechanisms to live disaster response applications, to achieve economic and social prosperity.

Chapter 8 reviews the major application of artificial intelligence and IoT in healthcare industry. The chapter also discussed the future scope of AI and IoT in healthcare and how this technology can be useful for medical practitioners.

Chapter 9 explores the integration of Digital Twin (DT) technology in transportation which enhances safety, efficiency, and decision-making in autonomous vehicles. The chapter discusses the technical challenges such as high implementation costs and data security concerns, benefits and future trends of DT technology that offers immense potential for sustainable urban transportation, smart city solutions, and advanced traffic management systems.

Chapter 10 examines the importance of augmented reality in medical teaching which allows students to learn better with specific links in the application, which permits interaction with 3D models of human anatomy.

The book highlights how innovative technologies such as AI, deep learning, IoT are reshaping industries, fostering innovation, and addressing key global challenges such as sustainability, resource efficiency, and personalized solutions. This book will serve as a valuable resource for researchers, practitioners, and undergraduate and post-graduate students who are eager to explore the transformative role of computer intelligence in the industrial revolution 5.0. By bridging the gap between theory and practical applications, this book will inspire further innovation and contribute to the advancement of intelligent, human-centered technologies.

Editors
Mukul Kumar Gupta
Abhinav Sharma
Vinay Rishiwal
Meera C S

Acknowledgements

The Editor acknowledges CRC Press - Taylor & Francis Group for this opportunity and professional support. My special thanks to Gagandeep Singh, Publisher Engineering, CRC Press, Gauravjeet Singh Reen, Senior Editor-Engineering and Mehnaz Hussain, Editorial Assistant for the excellent support provided us to complete this book.

Thanks to the chapter authors and reviewers for their availability for this work.

Chapter 1

Resnet50 architecture for multiclass brain tumor classification using fine-tuning technique

Vendra Durga Ratna Kumar, and Priyanka Singh

1.1 INTRODUCTION

Among the deadliest illnesses in the world are brain tumors, and they may affect any individual at any stage of life, gender, or ethnicity. An abnormal growth of tissue in the brain, which may or may not be malignant, is called a brain tumor [1]. Tumors related to Benign are not cancerous and come from brain-related tissues. Tumors like Malignant are rapidly expanding growths that have the potential to be deadly. Thus, early brain tumor detection is essential for improving patient outcomes [2]. The knowledge and experience of neurologists is essential in the traditional approach to a precise diagnosis of cancer. These days, brain-tumor cases are increasing rapidly, so researchers need to automate the diagnosis procedure. Figure 1.1 shows sample images from Kaggle MRI dataset. Neurologists benefit from computer-aided diagnostic systems in the following ways:

1. It reduces the requirement for human intervention and helps with early tumor detection for more effective treatment [3].
2. It gives the physician an additional perspective for making decisions [4].

Performing a classification of brain tumors is the most crucial step in the diagnosis process. This motivates scientists to implement a system for identifying brain diseases. The submitted brain image is classified into two categories: normal or abnormal, using the pathology detection technique. Nayak et al.'s [5] proposed rapid entropy features have been proposed for the diagnosis of brain diseases. Using transfer learning, Lu et al. [6] created pre-trained AlexNet to identify sick brains. Nayak et al. [7] have proposed a deep-learning neural network for brain-tumor diagnosis that is based on a sophisticated wavelet autoencoder. Elkorany et al. [8] used the parameter-free BAT optimization approach with Fisher criteria to categorize brain tumors. Vidyarthi et al. [9] have created an improved genetic algorithm designed to identify anomalies in the brain. Unlike binary classification, multi-class classification takes into account both the kind of brain tumor and the pathology detection. Out of the over 120 distinct types of

DOI: 10.1201/9781003568995-1

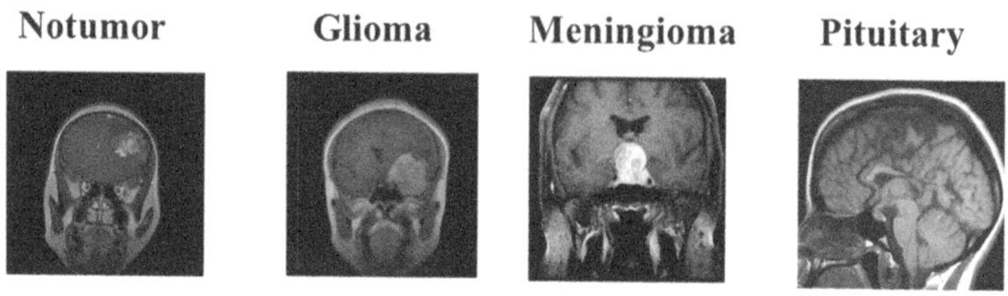

Figure 1.1 Images of different brain tumors and no tumor.

brain tumors, 21.5% are meningiomas, 42.11% are gliomas, and 7.89% are pituitary tumors [10]. In Figure 1.1, which displays various categories of MR (magnetic resonance) scans of each tumor, a green line designates the tumor area. Numerous characteristics, such as severity, surface, location, size, and shape, might indicate the kind of tumor. Figure 1.1(c) and (d) depict meningioma and pituitary tumors. The textures and intensities of these malignancies are consistent. These tumors differ, nevertheless, in terms of the tumor's shape, size, and location. Pituitary tumors and meningiomas form near the skull region with the pituitary gland, respectively. Glioma cancers impact glial tissue cells in the brain; It illustrates the range of intensity of these tumors [11]. A human specialist may easily identify a particular type of tumor by looking at these characteristics. Unfortunately, the range of features present in brain tumors makes it a difficult task for computer-aided tumor classification systems. This chapter uses residual networks combined with pre-training on a massive dataset called ImageNet, which has millions of different image categories [12], to understand lots of different features and patterns useful in tackling the multi-tumor classification issue.

The proposed model takes advantage of adjusting its architecture, making every layer trainable, and adding further layers on top of the foundational structure. Concurrently, strategic layer freezing is used to optimize the Resnet 50 basic architecture. To further assure optimum performance for both models and to help achieve the overall goal of improving brain-tumor classification accuracy, the hyper-parameters have been rigorously fine-tuned for brain tumor classification leveraging the power of deep learning to identify different types of brain tumors accurately [13]. To facilitate the learning process, used 'Adam' optimizer used [14], categorical cross-entropy, and sparse-categorical loss functions to handle the problems at hand. Our research model makes use of global average pooling to address the overfitting issue.

These issues are resolved by the residual network's (ResNet) architecture, which also provides a very efficient way of increasing recognition accuracy and finishing challenging jobs quickly. When it comes to fixing problems that came up during the deep CNN training process, ResNet is an effective method. ResNet50 is used to calculate the brain tumor detection and recovery technique's precision [15]. This article describes how the ResNet50 architecture which is shown in Figure 1.2 may be used to provide incredibly accurate results.

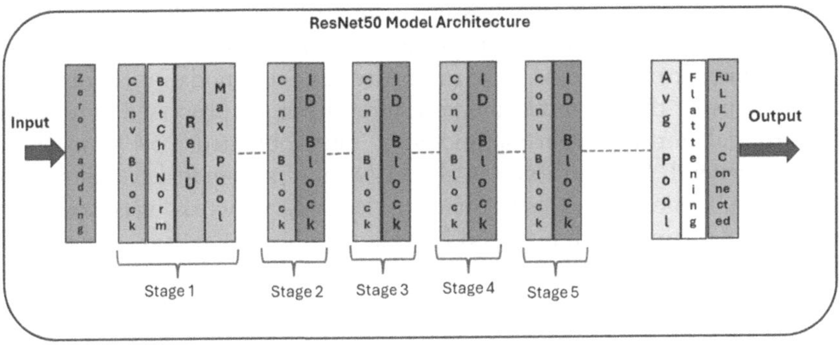

Figure 1.2 Architecture of the ResNet 50 base model [15].

Challenges that need to be considered for Obtaining extensive and meticulously annotated datasets to detect brain tumors presents difficulties since it is necessary to take into account patient privacy, a variety of representations, and detailed annotations. A lack of large-scale datasets with ground truth labeling might make it more difficult to create and validate reliable models.

1.2 RELATED WORKS

Gómez-Guzmán et al. [16] offered six pre-trained CNN models using the Fighshare dataset, which contains 7023 MRI images, In this dataset, there are four classifications of magnetic resonance imaging (MRI): one for notumor and three for brain malignancies, including pituitary, glioma, and meningioma. Xception, ResNet50, InceptionV3, InceptionResNetV2, MobileNetV2, and EfficientNetB0 are some of the preprocessing techniques used to train the models by using input MRI images. InceptionV3 attained an average accuracy of 97.12% when compared with other architectures. Praveen et al. [17] suggested SVMs, K-NN, multi-layer perceptrons, naive Bayes, and algorithms based on random forests and deep learning techniques for the classification and segmentation of brain tumors. The greatest classification accuracy of 92.4% was attained by classic SVMs out of all three methods. Saber et al. [18] implemented a customized CNN architecture and VGGNet to categorize 253 images of brain tumors, of which 155 were tumors and 98 were normal. On a specific dataset, VGGNet demonstrated a maximum validation accuracy of 98.96%, whilst the modified CNN model achieved a total validation accuracy of 86%. In this paper, Shah et al [19] authors examined Global thresholding, adaptive thresholding, Sobel filters, high-pass filters, median blurs, equalization of histograms, dilations, and erosions were all suggested by the authors. Furthermore, a pretrained ResNet101V2 model based on transfer learning was shown to identify brain tumors in 3762 images. They achieved an accuracy level of

95% in their findings. Similarly Khan et al. [20] proposed K-means clustering and the VGG19 CNN architecture for the segmentation and classification of brain cancers in MRI images. After an MR modality is converted to slices using the suggested method, intensity is preprocessed employing a statistical normalization methodology. They obtained a 94% accuracy rate in general. In Oberhofer et al. [21] a fully automated technique for classifying brain tumors using machine learning networks was presented to identify pictures with high and low glioma illness. The authors achieved 90% and 95% accuracy in multi-classifying brain cancers into primary, secondary, and neurological brain tumors using a severe gradient-boosting model. In [22], the researchers offered a variety of CNN-based classification algorithms, each with many repetitions, including the EfficientNetB0, VGGNet, GoogleNet, and ResNets. Compared to GoogleNet with 93.46% and VGGNet with 89.34%, EfficientNetB0 has a higher accuracy rates of 98.86%. Compared to VGGNet and GoogleNet, EfficientNetB0 generates results with 10% high accurate in less time. A deep artificial network-based method utilizing squeezing and excitation ResNet-101 was presented by Ghosal et al. [23] for the automatic categorization of brain tumors in magnetic resonance imaging.

The goal of the study was to distinguish between meningiomas, pituitary tumors, and gliomas. To boost performance, they used data augmentation and zero-centering intensity normalization. Comparing the experimental results to other recent approaches for classifying brain tumors, considerable improvements were observed in terms of precision, specificity, and sensitivity. A technique for classifying brain tumor images utilizing transfer learning along with fine-tuning has been put into practice by Swati et al. [24]. Using a five-fold cross-validation procedure, the authors' accuracy came in at 94.82%. Utilizing deep-transfer learning, Deepak et al. [25] have developed a three-class classification scheme to distinguish between brain tumors, glioma, and pituitary cancers. With 97.1% classification accuracy, a pretrained GoogleNet was utilized for extracting traits from brain MRI. To determine the excess season of amyotrophic sidelong sclerosis, researchers in [26] suggested an alternative method that relies on a DL-based method on clinical characteristics and an MRI scan of a sick patient. This method improves remaining time estimation by combining an MRI image and medical features into a modified CNN. ML systems are getting closer to AI thanks to deep artificial neural networks. Another CNN-based tumor treatment technique the cerebral MR image precision of categorization [27] is 97.18%. Another study classified tumors into four groups using MRI: gliomas, metastatic bronchogenic, sarcoma, and normal cancer tumors) utilizing a set of sixty-six brain MRIs [28]. Paut et al. [29] introduced a neural network modeling approach that leverages datasets for training across various categories. The best-performing neural network accomplished a standard fivefold cross validation accuracy of 91.43%. Details of the model's sequential layers, including output shapes and parameter counts is represented in Table 1.1.

Table 1.1 Explore an in-depth analysis of different approaches and methodologies for the classification of brain tumors

Author	*Methods used*	*Techniques*	*Dataset images*	*Purpose*	*Results (%)*
Gómez-Guzmán et al. [16]	InceptionV3	CNN	Br35H 7023	Detection of BT	97.12
Kumar et al. [17]	QSVM	Quantum machine learning model	Brats 2015	Classification	95
Saber et al. [18]	VGGNet	CNN	253	Classification	98.96
Shah et al. [19]	ResNet101V2	Transfer learning	3762	Classification	95
Khan et al. [20]	KMeans clusters algorithm, VGG19	CNN	3762	Classification	94
Oberhofer et al. [21]	Gradient boosting model	ML	3762	Multi-classification	95
Zulfiqar et al. [22]	EfficientNetB0	CNN	Kaggle 3000	Classification	98.86
Ghosal et al. [23]	ResNet-101	Artificial network-based	Kaggle 253	Multiclass classification	97.2
Swati et al. [24]	Transfer learning along with fine-tuning	Five-fold cross-validation	CE-MRI 3064	Multiclass classification	94.83
Deepak et al. [25]	Pretrained GoogleNet	Deep transfer learning	Figshare 6064	Three class classification	97.1
Vander Burgh et al. [26]	DL-based method	Deep artificial neural networks	135 Datasets	Classification	84.4
Ari et al. [27]	Tumor treatment technique	CNN-based	cerebral MR image	Classification	97.18
Paul et al. [29]	Dataset-based neural network	Neural network modeling technique using five-fold cross-validation	CE-MRI 3064	Classification	91.43

The table represents the performance and analysis of the suggested model by using a publicly accessible 4-class brain tumor dataset. As a result, current studies are concentrated and analyzed in the same dataset. The vast majority of these methods make use of deep neural networks. An issue known as the vanishing gradient affects deep neural networks. To solve the difficulty, Residual Network System (ResNet-50) and globally averaged pooling are utilized for various multiclass brain tumor classifications.

1.3 PROPOSED METHODOLOGY

This chapter describes our study's entire methodology. Our dataset is taken from a based computational platform, and it consists of 7023 MRI brain tumor and nontumor images. This dataset is partitioned into two sets. The first set consists of 5712 MR images used for training and the second dataset consists of 1311 MR images used for testing. From these two sets, four classifications are performed: normal cases, meningioma, glioma, and pituitary tumors. To ensure data consistency and quality, researchers implemented several pre- processing methods to resize every image to an acceptable size. Further preprocessed data is segmented with trainset and testset with an average ratio of 80% and 20%. The proposed ResNet50 neural network with many layers and pre-trained algorithms can be discussed further. Transfer learning with Finetuning based on ResNet50 using CNN Architecture for brain tumor image classification is performed.

1.4 DATA PREPROCESSING

Data preprocessing is a technique used to perform several important recipes for improved performance like image resizing, normalization, data augmentation, random chopping, color jittering, and data shuffling.

1. Image Resizing: Images are resized to a standard dimension that typically has 224×244 pixels, to generate a uniform input size for the ResNet50 model.
2. Normalization: Pixel values are scaled to a range of 0–1 or standardized to exhibit zero mean and unit variance. Normalization helps in stabilizing and speeding up the training process.
3. Data Augmentation: Several transformations are applied to improve the variety of the training dataset, such as rotation, flipping, zooming, flipping, and shearing.
4. Random Chopping: Randomly cropping a section of the image to match the required input size, helping the ResNet50 model to generalize better.

5. Color Jittering: Randomly altering the brightness, contrast, saturation, and hue of the images to make the ResNet50 more robust to color variations.
6. Data Shuffling: Randomly shuffling the dataset before each epoch to ensure that the model does not learn any ordering in the data and improves generalization.
7. Batch Normalization: Implementing batch normalization layers to stabilize and accelerate the training process by normalizing inputs in each batch.
8. Image Standardization: Transforming images into a standard format according to the dataset's mean and standard deviation, which were utilized to pre-train ResNet50.

The use of data pre-processing techniques enhances the performance of accuracy, generalization, and training efficiency on ResNet50 brain tumor classification, leading to more accurate and reliable predictions. Pre-processing techniques lead to faster convergence by reducing training time as model parameters are updated more effectively. This research took advantage of the freely accessible online Kaggle tool a database that depends on MRI images [29]. Three distinct testing, validation, and training set ratios were used for our tests: 80%, 10%, & 10%; 75%, 15%, & 10%; and 75%, 20%, and 5%. The best results were obtained with percentage combinations of 75%, 20%, and 5%. Before inserting the images into the prepared framework, the pre-processing stage is necessary to simplify computations in less time and with better speed, eliminating the numerical procedure and dimensions. The unorganized data is then divided into halves and used for training. This reduces the scope of the entire collection, saving time and effort. It contains authentic data that may be applied to training, validation, and testing. Table 1.2 uses enumerated parameters to illustrate each layer and output form of the ResNet50 model. The dimensions of size 224×224×3

Table 1.2 Input data processing matrix

Model: "Sequential" layer (type)	*Output shape*	*Number of parameters*
ResNet50 [Architecture]	(None, 7, 7, 2049)	3587712
Dropout Layer 3 [Dropout]	(None, 7, 7, 2049)	0
Flattern with 2 [Flatten]	(None, 100353)	0
Dropout Layer 4 [Dropout]	(None, 100353)	0
DenseNet with 2 [Dense]	(None, 1)	100354
Total parameters: 23, 688, 068		
Trained params: 100, 354		
Not Trained parameters: 23, 587, 715		

with pretrained resnet50 using CNN model architecture with multiple convolution operations using fully connected layer and SoftMax to classify tumors in the brain. These pre-processing techniques ensure that the input data is consistent and meaningful under the conditions of ResNet50 model was initially trained, facilitating effective transfer learning and fine-tuning.

This textual model summary contains details regarding:

- The model's layers and their arrangement.
- Each layer's final form.
- The quantity of weights, or parameters, in every layer.
- The total quantity of the model's weights, or parameters.

The summary() method on the model may be used to construct the summary; it produces a string used to represent output.

The ResNet50 models' implementation is given in this section. By designating all the neural network's layers as capable of training and adding them, our method entails fine-tuning the hyperparameters for both training and optimization. This allows us to modify the neural network's biases and learning rates during training. Adam optimizer helps the model to perform adaptive learning capabilities, memory efficiency, and fast convergence properties. Improving accuracy and decreasing total loss. Global Average Pooling, dropouts x2, dense layer using Softmax as a classifier. Below Figure 1.3 represents the block diagram of data preprocessing technique to the proposed methodology.

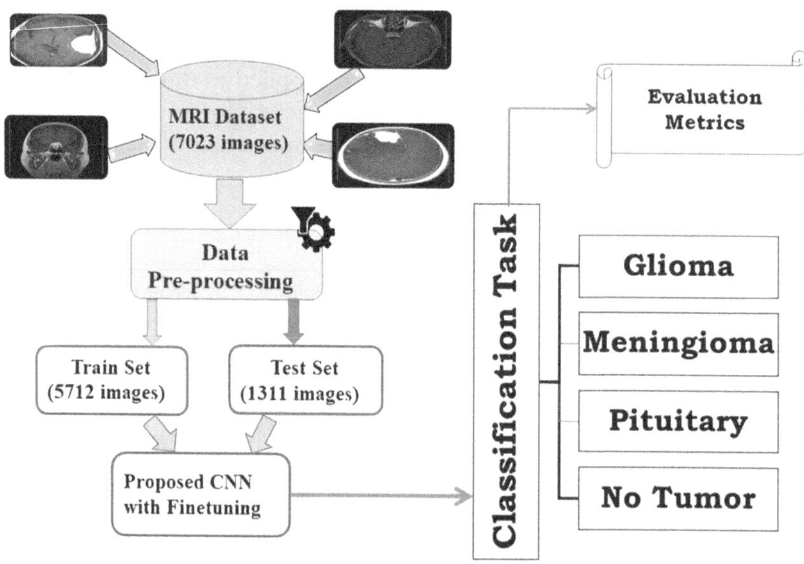

Figure 1.3 Proposed block diagram with finetuning.

The suggested method for detecting brain tumors uses four basic strategies: feature extraction, augmentation, transfer-learning, and fine-tuning techniques with the ResNet50 model using data pre-processing. Figure 1.3 depicts the proposed block diagram with fine tuning [28] design.

1.4.1 Architecture

The Recommended Architecture is shown in Figure 1.5. The process starts with the labeling of raw files and the import of certain photos. The enhancement approach is applied to data by splitting it into testing sets, validation sets and training sets. The recommended algorithm developed an optimum algorithm with regularization techniques and hyperparameters, suggesting a novel approach. The next sections will provide the network training results and the numerical efficiency of the suggested approach. The recommended structure process starts with the labelling of raw files and the import of certain photos. The enhancement approach is applied once the data is split into testing sets, verifying sets and training sets. Through the use of hyperparameters and regularization techniques, the recommended algorithm developed an optimum algorithm that offered a novel approach. The next sections will provide network training results and the numerical effectiveness of the suggested method.

1.4.2 Techniques for transfer learning

The proposed architecture takes advantage of transfer learning strategies to characterize models. It starts with an input layer consisting of a certain number of levels, each of which includes improved images from the previous data processing stage. Its activation functions cut across the convolutional layers. This technique uses downsampling to identify features. This phase prevents overfitting by using a dropout layer, a softmax Layer with a fully connected layer. Before forecasting class using classification layers, the outcome is computed. In the data normalization process, an input layer with confirmed picture resolution is employed [30]. A 2D convolution layer using kernel or convolution filtering of frame F and dimensions LM is implemented for the input images. The filters are used to keep track of the inputs and use the dots that are the result of the inputs to determine the weights of the kernel.

Moving filters across a picture is known as stride (S), and it works similarly to vertical and horizontal steps. The data around the edges of the first images that were taken right before moving are kept by padding (P). At the lowest level, these kernels are feature markers that are limited to identifying edges, lines, and bumps. Conversely, higher learning levels of learning strategies are employed to recognize more complex traits. The proposed work function block diagram is demonstrated in Figure 1.4.

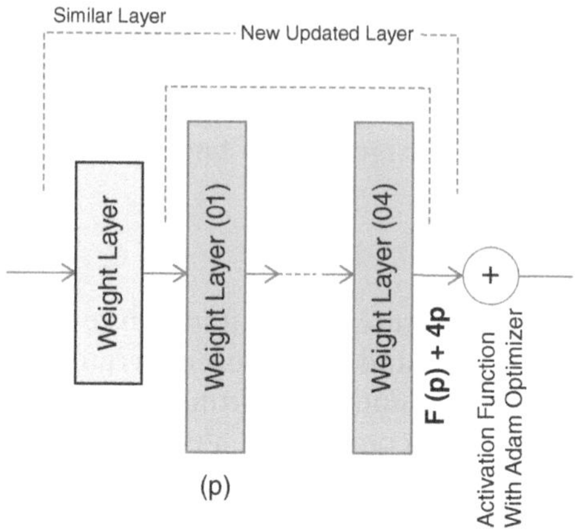

Figure 1.4 An example of how the block function is used [31].

The suggested algorithm's block diagram indicates adjusted the projected model (ResNet50) by the transfer learning. As a result, it has been incorporated into the primary component diagram along with layer updates and function additions following the modeling phase. Thus, by the suggested approach data flow diagram, both are primary functional diagrams.

The suggested approach uses the idea of transfer learning, which avoids replicating the network. This idea states that, depending on the demands of the target model, the original model only has to be slightly modified.

1.4.3 Dataset

Four various brain tumors types are - Meningioma, Pituitary, Glioma, and Notumor—are represented in a broad "Brain Tumor Classification" dataset that includes 7,023 images obtained from Kaggle. A split of dataset was made, with 80% going to train the dataset and 20% going to test the dataset. To create and train the networks, training dataset is used. This dataset offers a thorough basis for assessing ResNet50's performance in a multiple-class classification system.

1.4.4 Proposed Resnet50 architecture

ResNet50 is a residual neural network with 50 levels and multiple sets of comparable layers. For instance, ResNet50 is capable of managing neural network training dataset degradation problems, rising slopes, and the false issue of disappearing borders. In 2016, He et al. proposed the first ResNet technique [32]. Because ResNet101 and ResNet50 are so easy to use and beneficial, several alternative ways have been created. In various domains, such as object identification, segmentation, and detection, several experts now favor the ResNet approach. The accuracy of training data collection decreases as the neural network depth increases [33, 34]. Since an overfitting scenario necessitates great precision, knowing the true cause is essential. He et al. [32] therefore proposed that a deep residue network allows for detailed network tweaking.

To minimize overfitting, attain maximum efficiency in identifying patterns in the data, and provide efficacy, the ResNet50 design uses skip connections. By the network's architecture, ResNet50 has 50 layers and an input shape of 224 × 224. Its ability to identify patterns in the information makes the model effective. Many unique patterns and traits that are important to our categorization job are made easier to grasp with the support of ResNet50 (https://towardsdatascience.com/the-annotated-resnet-50-a6c536034758).

$$Y = F(x) + x$$

where Y is the residual function's output parameter, x is its input, and () is residual function. Figure 1.5 displays the architecture of the ResNet50.

ResNet50 comprises 50 layers, hence its name, and is structured around residual blocks. The core idea behind ResNet50 is outline of skip connections or shortcuts that make gradients to flow more directly through the network during training. This mitigates a vanishing gradient problem that enables training of deeper networks. The residual block, which comprises of many layers of convolution followed by skip connections, is the fundamental component of ResNet50. The representation of each block is as follows:

Convolutional Layers: Typically, each residual block in ResNet50 contains multiple convolutional layers, often with batch normalization and ReLU activation functions.

Shortcut Connection: The shortcut connection bypasses one or more convolutional layers and directly adds the input to the output of the block. This enables the network to learn residual functions, making it easier to optimize deeper networks.

Identity Mapping: If the dimensions of input and output of residual block are the same, the shortcut connection directly adds input to the output.

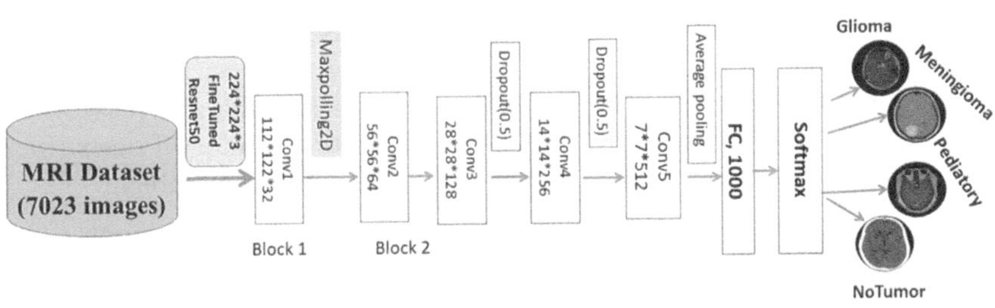

Figure 1.5 ResNet50 proposed model architecture.

If the dimensions differ, the input goes through a 1x1 convolutional layer to match the dimensions of the output before being added.

Maxpooling: Maxpooling is a downsampling operation commonly used in CNN architectures. It partitions the input image into non-overlapping rectangular regions and outputs the maximum value from each region. Maxpooling helps in dropping the spatial dimensions of feature maps, leading to computational efficiency and translational invariance.

In ResNet50, Maxpooling layers are strategically placed at certain intervals to downsample feature maps and by capturing the most salient information from the input.

1.4.5 ResNet50 with average pooling and dropouts for brain tumor classification

When ResNet50 is employed for brain tumor classification, replacing Maxpooling with AveragePooling and incorporating dropout layers can enhance its performance:

AveragePooling: Unlike Maxpooling, which takes the maximum value from each region, AveragePooling calculates the average value. This can help in capturing more nuanced features from the data.

Dropout: During training, a portion of neurons are randomly removed as part of the dropout regularization approach. Making the network learn redundant representations, helps avoid overfitting and improves the network's capacity for generalization.

1.4.6 Contribution to enhanced brain tumor classification

Feature Representation: The deep architecture of ResNet50 allows it to automatically learn hierarchical representations of features from brain tumor images by capturing low-level details with high-level abstract patterns.

Residual Connections: The residual connections enable the network to effectively propagate gradients, facilitating the training of deeper networks without suffering from the vanishing gradient problem.

$$Y = F(p) + p, \text{ make } F(p) = 0 \text{ to make } Y = p$$

$F(p)$ — Calculating the loss function.

The logic behind Residual Networks is to make $(Y = p)$, where p is input, Y is output, $Y = F(p)$ in traditional neural networks.

Input is used to calculate the loss function, and the result is Y. In conventional networks, the output value is computed after the input passes through each layer. The above Figure 1.6 illustrates how skip connections are utilized to add the real input to the output. The network learns from the variance between input and output to raise overall accuracy.

Creating $F(p) = 0$ makes it simple to set the input and output equal.

$$Y = p + F(p)$$
$$Y = p + 0$$
$$Y = p$$

Unlike normal networks, that learn from Y, residual networks teach us using $F(p)$, and our goal is to attain $F(p) = 0$. Then only input and output are made equal.

AveragePooling: By replacing Maxpooling with AveragePooling, the network can capture more nuanced information from the images, potentially improving classification accuracy. **Dropout Regularization**: Dropout helps prevent overfitting, allowing the model to simplify better to hidden data and improving its robustness in classifying brain tumor images.

The ResNet50 models' implementation is detailed in this section by designating all the neural network's layers as capable of training and adding them, our method entails fine-tuning the hyperparameters for both training and optimization. The performance of ResNe50 allows us to modify the neural network's biases and learning rates while improving accuracy and decreasing total loss. Global Average Pooling, dropouts x2, dense layer using Softmax as a classifier. The Better ResNet50 Model is produced by removing the final layer from the ResNet50 and including the dropout, flat, network, residual, and thick layers. The probability value can be handled and understood more easily than the model's output, which is why the sigmoid function follows. The ADAM Optimizer was incorporated into the suggested model as well, as its enhanced gradient allowed it to perform far better than the previous optimizers.

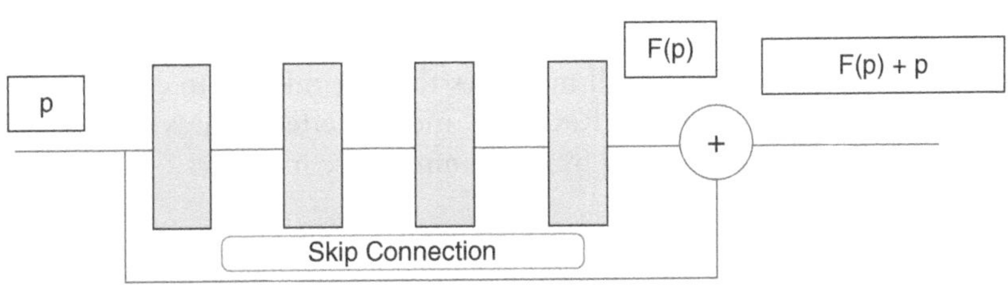

Figure 1.6 Residue block with skip connection.

1.5 LIMITATIONS

Although ResNet50 is a strong deep-learning architecture for a variety of computer vision applications, it has many drawbacks. ResNet50 requires a lot of processing resources to train and deploy because of its high number of parameters. Overfitting is a common problem with ResNet50, particularly when training on small datasets. Interpretability in medical diagnosis is lacking in ResNet50. The selection of hyperparameters may have an impact on ResNet50's performance. Perhaps customized models could provide a better fit. Compared to VGG16, InceptionV3, Xception, and EfficientNetB0 While ResNet50 is still a robust and adaptable model, newer models such as EfficientNetB0, Xception, and InceptionV3 are more appealing for different applications due to their limits in terms of efficiency, model size, and occasionally accuracy.

1.6 TRAINING

A brain tumor dataset was used to train the pre-trained ResNet-50. This procedure modifies the model weights to make sure that the final model is capable of classifying brain tumor images into four groups. For quicker convergence, a randomized gradient descent with momentum optimizer is used during training. Additionally, it lessens the computational load on applications with big data. Lastly, the suggested RNGAP model is used to test data for brain tumor classification.

1.7 PERFORMANCE EVALUATION METRICS

To analyze and assess the performance of the suggested model, Five-Fold cross validation technique is employed. The model is trained using 5712 images from the Kaggle dataset, and it is tested with 1311 images. Imagenet weights are used to train the Resnet50 model. The maxpolling technique is utilized to achieve 175 model layers, and batch normalization with 32 batch sizes and a total of fifteen epochs yielded an accuracy of 99.85% with a loss of 0.41%, a validation accuracy of 99.39%, and a validation loss of 4.60. Table 1.3 also shows that the suggested model consistently performs at an average of 0.99 across all measures that are taken into consideration. A thorough evaluation of the ResNet50 model's effectiveness in classifying brain tumors may be obtained by combining these measures.

Accuracy:

Accuracy represents the percentage of cases that were accurately predicted divided by total predicted.

Table 1.3 ResNet50 architecture performance evaluation and other deep learning models

Architecture name	*Accuracy%*	*Sensitivity%*	*Specificity*	*F1-score %*
VGG16	98.51	99.0	98.4	99.0
InceptionV3	97.52	97.6	97.7	96.4
Xception	97.84	98.5	96.6	97.5
EfficientNetB0	98.83	99.5	99.4	98.9
ResNet50	97.61	98.2	97.6	97.9
Pre-trained ResNet50	99.39	99.25	99.25	99.5

$$\text{Accuracy} = \frac{\text{TP} + \text{TN}}{\text{TP} + \text{TN} + \text{FP} + \text{FN}}$$

The accuracy usage gives a broad indication of the model's accuracy rate.

Precision:
Precision gives a percentage of all expected positives to accurately predict positive observations.

$$\text{Precision} = \frac{\text{TP}}{\text{TP} + \text{FP}}$$

Precision usage is to accurate the positive predictions, which is crucial for reducing false positives in medical diagnosis.

Sensitivity/Recall:
Sensitivity shown the proportion of all observations made in the actual class to all properly anticipated positive observations.

$$\text{Recall} = \frac{\text{TP}}{\text{TP} + \text{FN}}$$

Its usage is to evaluate the model strength to identify positive cases, which is essential for making sure brain tumor patients are not overlooked.

F1 Score:
F1 score gives harmonic mean of precision and recall.

$$\text{F1 score} = 2 \times \frac{\text{Precision} + \text{Recall}}{\text{Precision} \times \text{Recall}}$$

F1 score usage is to balances recall and accuracy, which is helpful in situations for distribution of classes that is not uniform.

Specificity:

Specificity defines the proportion of all observations in the actual negative class to all properly predicted negative observations.

$$\text{Specificity} = \frac{\text{TN}}{\text{TN} + \text{FP}}$$

Its usage is to evaluate the model's accuracy in identifying negative instances, which is crucial for comprehending false positive rates.

Area under the curve - Receiver Operating Characteristic curve (AUC-ROC):

- **AUC-ROC defines** the genuine positive rate vs the rate of false positives shown against the area beneath the ROC curve.
- Gives a total performance indicator for all categorization thresholds, which is helpful for comparing models.

Confusion Matrix:

- Confusion Matrix shows a table that displays a performance of a classification model on a set of test data with known real values.
- Provides a thorough overview of the model performance by displaying the quantity of actual positives, actual negatives, false positives, and incorrect results.

The metrics selected will rely on the particular requirements of the process of classifying brain tumors for an in-depth analysis. A thorough evaluation of the ResNet50 model's effectiveness in classifying brain tumors may be obtained by combining these measures. Figure 1.7 shows the confusion matrix of the proposed ReseNet50 architecture.

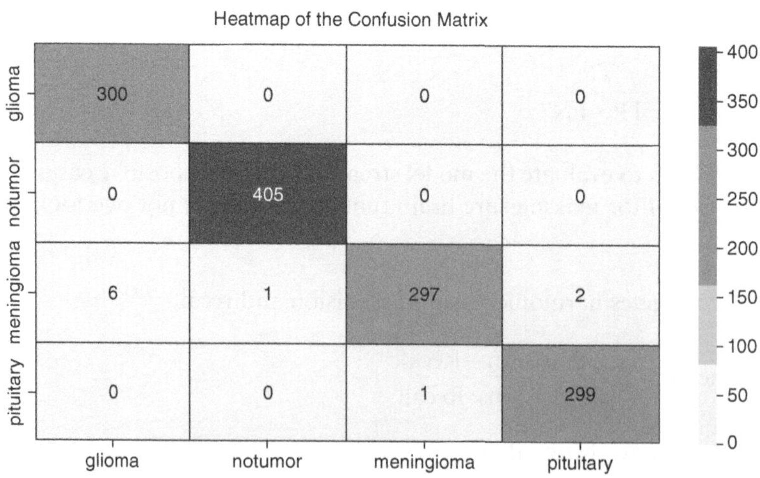

Figure 1.7 ResNet50 confusion matrix.

1.8 COMPARATIVE STUDY

Ma et al. [35] show the variations in different segmentation and detection techniques on the PHE-SICHCT-IDS by conducting several trials with both deep learning and standard machine learning techniques. Semantic segmentation achieves the highest precision of 76.31% while identifying objects reaches the best precision of 97.62%. The results of the MRI feature collection and analysis experiment show that the recommended features are useful in predicting SICH patients' diagnoses and that PHE-SICH-CT-IDS is a useful tool for evaluating image features.

Jain et al. [36] suggested multi-modal for the ongoing study utilizing several architectures, such as Inception V3 modules, CNN, VGG16, and ResNet50. After analyzing the work completed, it was found that ResNet50 (Fine Tuning) produced the best overall results with 98.63%, 97% in the detection of GBM tumors, 98% in the detection of Meningiomas (MEG), 98% in the detection of Pituitary tumors (PITT), and 99% in the detection of No Tumors (NT). The examined results on a dataset consisting of 7022 MRI images. The resilience of suggested ResNet50 (fine-tuning) design is demonstrated by the experimental findings.

Mundada et al. [37] proposed a study to assess and contrast the effectiveness of many deep learning algorithms that may be applied to the categorization of brain tumors. The experimental study makes use of six pre-trained TL classifiers. Brain tumors are automatically identified and classified using InceptionV3, Xception, Res net50, MobileNet, VGG16, and EfficientNetB0 using a fine-grained categorization method. This experiment employs an abnormal brain MR imaging dataset containing 7022 publicly available images on Kaggle, and Python is the software used. A frequently used matrix is utilized to compute the results. According to classification studies, this VGG16 TL algorithm performed exceptionally well and achieved the greatest accuracy of 99.09%.

Hussen et al. [38] incorporate four conventional classifiers (support vector machines (SVM), Naive Bayes Machine (NB), Decision trees (DT), and random forests (RF) with a hybrid model (VGG16). The experiment results show that 99.17% is the best performance for VGG16 obtained when Random Forest is combined with region-based segmentation.

Yalamanchili et al. [39] proposed a Visual Geometry Collective with 16 parameters determined for training the brain tumor detection model. VGG-16 and Efficient NetB7 are two of the best-performing simple convolutional neural network (CNN) models to perform well. The study's findings show the accurate, timely, and efficient use of MRI in the detection of brain cancers. To train and evaluate this model, 7022 brain MRI images were employed. Empirical results show that the proposed asymmetrical deep CNN model, which integrates aberrant and conventional imagery, is 98.19% successful in brain tumor classification using MRI images.

With the study of literature review it is confirmed that ResNet50 is a robust model with strong training stability and good accuracy. Its training times are greater than those of some other models, and it requires a significant amount of processing power. Its shortcomings advise looking at other models when computing efficiency is a top concern. On the other hand, its virtues make it appropriate for applications demanding great precision.

1.9 DISCUSSION

The proposed model has performed pre-processing, architectural reconfiguration, and fine-tuning by adding a few more layers to our proposal to create a unique model that can effectively identify brain cancers. When compared to other previous studies, our effective classification of brain tumors performed much better in terms of accuracy. The comparative analysis of brain tumor categorization is displayed in Table 1.4. Our solution yields the highest accuracy among all the measured performances while using the identical [40] dataset and the same amount of photos (7022). An equitable comparison to estimate efficiency is made possible by the usage of an identical dataset and quantity of photos.

The study shows the highest accuracy rate of 99.68 percent for a brain tumor disease dataset by utilizing RestNet50, which performs much better than other methods.

Table 1.4 Comparative analysis of brain tumor classification using Kaggle and Figshare datasets

Sl.No	*Author*	*Architecture*	*Technique*	*Dataset*	*Number of images*	*Accuracy (%)*
1	Ma et al. [35]	PHE-SICH-CT-IDS	R-CNN	Figshare	7022	97.62
2.	Jain et al. [36]	ResNet50	Fine tunning	Kaggle	7022	98.63
3.	Mundada et al. [37]	VGG16	Transfer learning	Kaggle	7022	99.09
4.	Hussen [38]	VGG16	pre-trained TL	Kaggle	7022	99.9
5.	Yalamanchili [39]	deep CNN model	EfficientnetB7	Kaggle	7022	98.19
6.	Proposed Model	ResNet50	CNN	Kaggle	7022	99.39

1.10 RESULTS AND ANALYSIS

The results show that ResNet50 outperforms VGG16, InceptionV3, Xception, and EfficientNetB0 in the classification of brain tumors, achieving a 99.39% accuracy. This difference in performance highlights how crucial it is to choose the right deep-learning architecture. The precise outcomes of ResNet50 can be attributed to its effective architecture, scalability, and excellent transfer learning. ResNet50 model architecture is performed with pre-processing, architectural reconfiguration, and fine-tuning by adding a few more layers to our proposal to create a unique model that can effectively identify brain cancers. When compared to other previous studies, our effective brain tumor classification approach demonstrated a markedly improved level of accuracy. The comparative analysis of brain tumor categorization is displayed in Table 1.4. The model selection for brain tumor classification tends to be contingent upon the particular demands of the work, such as the simplicity of implementation, the required level of accuracy, and the available computer resources, as shown by its strengths and limitations.

Proposed ResNet50 solution offers the highest accuracy performance across all precision performances using the same dataset [40] and the same quantity of images (7022). An equitable comparison to estimate efficiency is made possible by the usage of an identical dataset and quantity of images. Figures 1.8 and 1.9 shows the training and validation accuracy and loss of the proposed model. It shows that the maximum accuracy rate of 99.39% for a brain tumor dataset is being utilized for RestNet50, which performs much better than other methods.

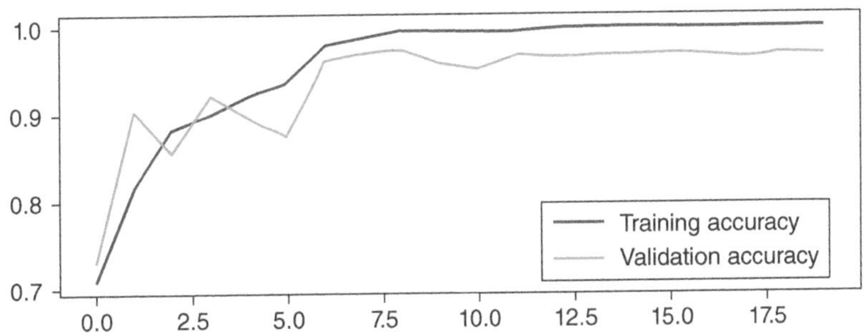

Figure 1.8 Accuracy in training and validation.

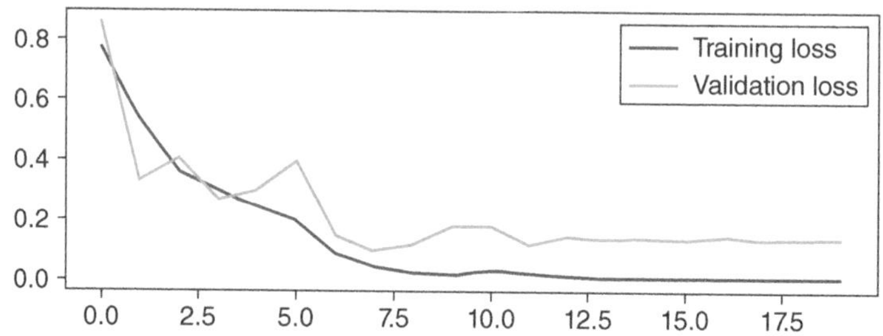

Figure 1.9 Loss in training and validation.

1.11 CONCLUSION AND FUTURE WORK

This research demonstrates that performing Fine-Tuning with Transfer Learning to the ResNet50 architecture predicts brain tumors into several classes using the magnetic resonance imaging image dataset. ResNet50 model beats the current deep neural networks frameworks in the multiclass brain tumor categorization task with an appropriate accuracy of 99.39%. This work offers various intriguing avenues for categorization of brain tumor in the near future. It is worthwhile to investigate the practical application of ResNet50 in medical frameworks for standard brain tumor evaluation and treatment planning. It is possible to develop a proficient deep learning-based framework for identifying the various stages of brain tumors.

REFERENCE

[1] Arumugam, M., Thiyagarajan, A., Adhi, L., & Alagar, S. (2024). Crossover smell agent optimized multilayer perceptron for precise brain tumor classification on MRI images. *Expert Systems With Applications*, *238*, 121453.
[2] Nadeem, M. W., Ghamdi, M. A. A., Hussain, M., Khan, M. A., Khan, K. M., Almotiri, S. H., & Butt, S. A. (2020). Brain tumor analysis empowered with deep learning: A review, taxonomy, and future challenges. *Brain Sciences*, *10*(2), 118.
[3] Ural, B., Özışık, P., & Hardalaç, F. (2020). An improved computer-based diagnosis system for early detection of abnormal lesions in the brain tissues using magnetic resonance and computerized tomography images. *Multimedia Tools and Applications*, *79*(21), 15613–15634.
[4] Windisch, P., Weber, P., Fürweger, C., Ehret, F., Kufeld, M., Zwahlen, D., & Muacevic, A. (2020). Implementation of model explainability for a basic brain tumor detection using convolutional neural networks on MRI slices. *Neuroradiology*, *62*, 1515–1518.
[5] Nayak, D. R., Dash, R., & Majhi, B. (2020). Automated diagnosis of multiclass brain abnormalities using MRI images: A deep convolutional neural network based method. *Pattern Recognition Letters*, *138*, 385–391.

[6] Lu, S., Wang, S. H., & Zhang, Y. D. (2020). Detecting pathological brain via ResNet and randomized neural networks. *Heliyon*, *6*(12), e05625.

[7] Nayak, D. R., Padhy, N., Mallick, P. K., & Singh, A. (2022). A deep autoencoder approach for detection of brain tumor images. *Computers and Electrical Engineering*, *102*, 108238.

[8] Elkorany, A. S., & Elsharkawy, Z. F. (2020). Automated optimized classification techniques for magnetic resonance brain images. *Multimedia Tools and Applications*, *79*(37), 27791–27814.

[9] Swati, Z. N. K., Zhao, Q., Kabir, M., Ali, F., Ali, Z., Ahmed, S., & Lu, J. (2019). Brain tumor classification for MR images using transfer learning and fine-tuning. *Computerized Medical Imaging and Graphics*, *75*, 34–46.

[10] Shah, A., Ishfaq, A., Malik, M. J., Shah, S. M. A. A., Nasir, T. B., & Fatima, S. R. (2022). Experience of brain tumors in tertiary care hospital. *Pakistan Journal of Medical & Health Sciences*, *16*(06), 149–149.

[11] Strong, M. J., Garces, J., Vera, J. C., Mathkour, M., Emerson, N., & Ware, M. L. (2015). Brain tumors: Epidemiology and current trends in treatment. *Brain Tumors Neurooncol*, *1*(1), 1–21.

[12] Nazir, M., Shakil, S., & Khurshid, K. (2021). Role of deep learning in brain tumor detection and classification (2015 to 2020): A review. *Computerized Medical Imaging and Graphics*, *91*, 101940.

[13] Wambugu, J. K. (2022). *A transfer learning and two-level hyperparameter optimization based model for improved classification of diabetic retinopathy*. Doctoral dissertation.

[14] Mehmood, F., Ahmad, S., & Whangbo, T. K. (2023). An efficient optimization technique for training deep neural networks. *Mathematics*, *11*(6), 1360.

[15] Feng, V. (2017). An overview of ResNet and its variants. *Towards Data Science*, *2*.

[16] Gómez-Guzmán, M. A., Jiménez-Beristaín, L., García-Guerrero, E. E., López-Bonilla, O. R., Tamayo-Perez, U. J., Esqueda-Elizondo, J. J., ... & Inzunza-González, E. (2023). Classifying brain tumors on magnetic resonance imaging by using convolutional neural networks. *Electronics*, *12*(4), 955.

[17] Kumar, T., Kumar, D., & Singh, G. (2023). Brain tumor classification using quantum support vector machine learning algorithm. *IETE Journal of Research*, 70, 4815–4828.

[18] Saber, A., Sakr, M., Abo-Seida, O. M., Keshk, A., & Chen, H. (2021). A novel deep-learning model for automatic detection and classification of breast cancer using the transfer-learning technique. *IEEE Access*, *9*, 71194–71209.

[19] Shah, H. A., Saeed, F., Yun, S., Park, J. H., Paul, A., & Kang, J. M. (2022). A robust approach for brain tumor detection in magnetic resonance images using finetuned EfficientNet. *IEEE Access*, *10*, 65426–65438.

[20] Khan, O. I. (2020). A hybrid unsupervised classification technique for automatic brain MRI tumor recognition. Doctoral dissertation. University of Guelph.

[21] Oberhofer, A., Bronkhorst, A. J., Uhlig, C., Ungerer, V., & Holdenrieder, S. (2022). Tracing the origin of cell-free DNA molecules through tissue-specific epigenetic signatures. *Diagnostics*, *12*(8), 1834.

[22] Zulfiqar, F., Bajwa, U. I., & Mehmood, Y. (2023). Multi-class classification of brain tumor types from MR images using EfficientNets. *Biomedical Signal Processing and Control*, *84*, 104777.

[23] Ghosal, P., Nandanwar, L., Kanchan, S., Bhadra, A., Chakraborty, J., & Nandi, D. (2019, February). Brain tumor classification using ResNet-101 based squeeze and excitation deep neural network. In *2019 Second International Conference on Advanced Computational and Communication Paradigms (ICACCP)* (pp. 1–6). IEEE.

[24] Swati ZNK, Zhao Q, Kabir M, Ali F, Ali Z, Ahmed S, Lu J (2019) Brain tumor classification for MR images using transfer learning and fine-tuning. *Comput Med Imaging Graph*, *75*, 34–46.

[25] Deepak, S., & Ameer, P. M. (2019). Brain tumor classification using deep CNN features via transfer learning. *Computers in Biology and Medicine*, 111, 103345.

[26] van der Burgh, H. K., Schmidt, R., Westeneng, H. J., de Reus, M. A., van den Berg, L. H., & van den Heuvel, M. P. (2017). Deep learning predictions of survival based on MRI in amyotrophic lateral sclerosis. *NeuroImage: Clinical*, *13*, 361–369.

[27] Ari, A., & Hanbay, D. (2018). Deep learning-based brain tumor classification and detection system. *Turkish Journal of Electrical Engineering and Computer Sciences*, *26*(5), 2275–2286.

[28] Mohsen, H., El-Dahshan, E. S. A., El-Horbaty, E. S. M., & Salem, A. B. M. (2018). Classification using deep learning neural networks for brain tumors. *Future Computing and Informatics Journal*, *3*(1), 68–71.

[29] Paul, J. S., Plassard, A. J., Landman, B. A., & Fabbri, D. (2017, March). Deep learning for brain tumor classification. In *Medical Imaging 2017: Biomedical Applications in Molecular, Structural, and Functional Imaging* (Vol. 10137, pp. 253–268). SPIE.

[30] He, K., Zhang, X., Ren, S., & Sun, J. (2016). Deep residual learning for image recognition. In *Proceedings of the IEEE conference on computer vision and pattern recognition* (pp. 770–778).

[31] Sharma, A. K., Nandal, A., Dhaka, A., Zhou, L., Alhudhaif, A., Alenezi, F., & Polat, K. (2023). Brain tumor classification using the modified ResNet50 model based on transfer learning. *Biomedical Signal Processing and Control*, *86*, 105299.

[32] Mohsen, H., El-Dahshan, E. S. A., El-Horbaty, E. S. M., & Salem, A. B. M. (2018). Classification using deep learning neural networks for brain tumors. *Future Computing and Informatics Journal*, *3*(1), 68–71.

[33] Nair, V., & Hinton, G. E. (2010). Rectified linear units improve restricted Boltzmann machines. In *Proceedings of the 27th International Conference on Machine Learning (ICML-10)* (pp. 807–814).

[34] Civilibal, S., Cevik, K. K., & Bozkurt, A. (2023). A deep learning approach for automatic detection, segmentation, and classification of breast lesions from thermal images. *Expert Systems with Applications*, *212*, 118774.

[35] Ma, D., Li, C., Du, T., Qiao, L., Tang, D., Ma, Z., … & Sun, H. (2024). PHE-SICH-CT-IDS: A benchmark CT image dataset for evaluation semantic segmentation, object detection and radiomic feature extraction of perihematomal edema in spontaneous intracerebral hemorrhage. *Computers in Biology and Medicine*, 108342.

[36] Jain, P., & Santhanalakshmi, S. (2022, October). Early detection of brain tumor and survival prediction using deep learning and an ensemble learning from radiomics images. In *2022 IEEE 3rd Global Conference for Advancement in Technology (GCAT)* (pp. 1–9). IEEE.

[37] Mundada, K., Rahatwilkar, T., & Kulkarni, J. (2022, September). Experimental analysis of deep learning algorithms used in brain tumor classification. In *Congress on intelligent systems* (pp. 91–100). Springer Nature Singapore.

[38] Hussein, A. M., Sharifai, A. G., Alia, O. M. D., Abualigah, L., Almotairi, K. H., Abujayyab, S. K., & Gandomi, A. H. (2024). Auto-detection of the coronavirus disease by using deep convolutional neural networks and X-ray photographs. *Scientific reports*, *14*(1), 534.

[39] Yalamanchili, S., Yenuga, P., Burla, N., Jonnadula, H., & Bolem, S. C. (2024). MRI brain tumor analysis on improved VGG-16 and efficient NetB7 models. *Journal of Image and Graphics*, *12*(1), 103-116.

[40] Dataset Link https://www.kaggle.com/datasets/masoudnickparvar/brain-tumor-mri-dataset

Chapter 2

Reinforcement learning for decision making

Nazeerunnisa Qureshey, Nandini Prasad K. S., Nisha S. L., Yarlagadda Anuradha, Chaitanya Krishna Kasaraneni, and M. Nagabhushana Rao

2.1 INTRODUCTION

Reinforcement Learning (RL) has shown itself to be a powerful framework in respect of decision-making, where an agent interacts with environment towards maximizing cumulative rewards [1]. Unlike traditional supervised learning methods, which rely on labeled datasets, RL is characterized by its trial-and-error learning process. This chapter explores fundamental concepts of RL, its components, popular algorithms, and its applications in various domains. Figure 2.1 explains reinforcement learning process. The diagram consists of Environment, Agent, State, Rewards, Action, value Function, Policy and Q-Function.

Agent is learner or decision-maker which engages with its surroundings. Environment is the external system the agent interacts with, that consists of states, actions, and rewards. State represents the current condition or situation within environment. An action is one of the options agent can choose, influencing state in some way. Reward is the numerical feedback signal the agent receives post performing action in particular state, which supplements guide to its learning process.

A policy is the approach or strategy that directs agent's actions based on different states. Value function predicts the total reward the agent can expect from certain state if it follows a particular policy. Q-function offers an estimate of the total expected reward for taking specific action in given state. RL process can be enumerated as follows.

a. Current state s_t of environment is observed by agent.
b. Action based on its policy π is selected by agent.
c. Environment responds to action, transitioning to new state s_{t+1} and providing reward r_t
d. Based on received reward and new state, agent updates its knowledge in form of policy or value function.

As loop continues, agent keep learning from its experiences over time [2]. Mainly there are two approaches, which are as follows.

DOI: 10.1201/9781003568995-2

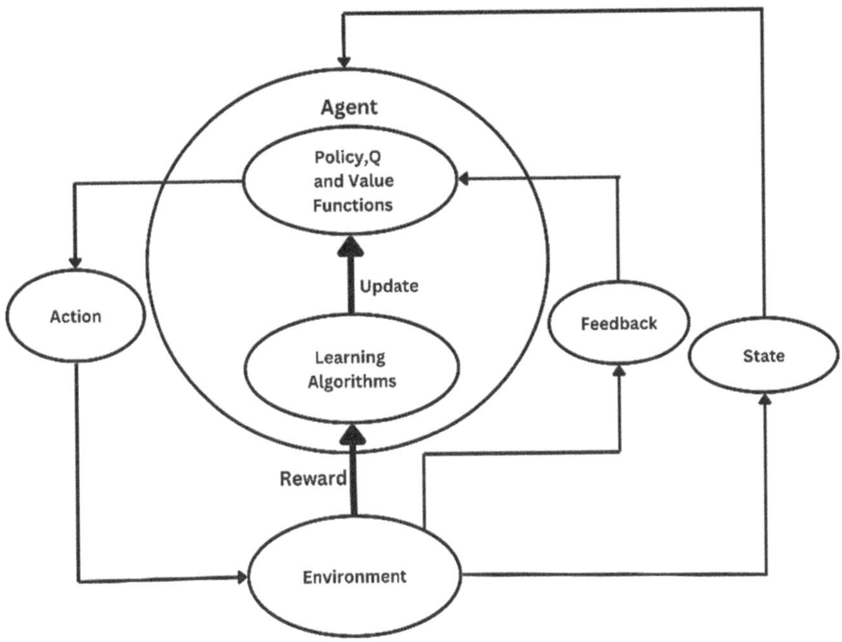

Figure 2.1 Overview of reinforcement learning.

2.2 MODEL-FREE METHODS

Model-free approaches allow agents to learn policies or value functions directly by interacting with environment, bypassing the need to construct a model of its underlying dynamics [3]. These methods offer reliable frameworks for discovering optimal strategies across diverse settings, even when there is no prior knowledge of environment's structure. Such techniques learn from their surroundings, supporting applications that range from gaming to robotics and more. The overview of Model Free structure has been shown in Figure 2.2. As the field continues to evolve, these methods will remain central to advancing capabilities of intelligent systems. Flow of model-free RL algorithm is represented as follows.

a. **Initialize.** Start with initial policy.
b. **Observe State.** Observe current state.
c. **Select Action.** Use exploration strategy to select action.
d. **Observe Reward.** Execute action, receive reward, and observe new state.
e. **Update.** Update policy using observed reward and state transition.
f. **Repeat.** Repeat process till policy converges.

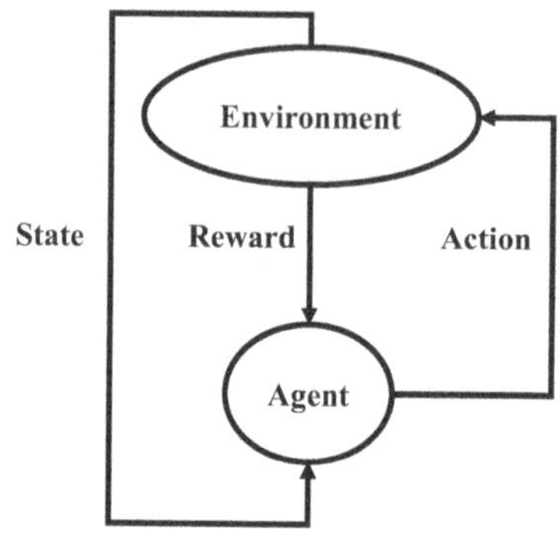

Figure 2.2 Overview of model free structure.

This flow continues as agent learns from interactions and gradually improves its decision-making policy. In subsequent paragraphs, we will see some of key model-free methods, highlighting their mechanisms and applications.

2.2.1 Q-learning

It is an algorithm that finds optimal action-selection policy for a particular Markov Decision Process (MDP) [4]. It does so through learning an action-value function, $Q(s,a)$, that represents the expected return (or cumulative reward) of taking action '*a*' in state '*s*' and following the optimal policy subsequently. Q-learning is an off-policy algorithm, meaning that it learns optimal policy independently of actions taken by agent during learning (it does not need to follow current policy to learn optimal behaviour). In Q-learning, the goal of the agent is to optimize cumulative future reward, known as return or expected reward. To achieve this, the agent seeks to discover policy that optimizes its overall reward and it interacts with environment through following components.

a. **States (S)**. These indicate various situations or configurations, agent may be in, within environment.
b. **Actions (A)**. These are possible actions agent can take in each state. Each action results in a transition from one state to other.
c. **Rewards (R)**. Reward is a value, agent takes after performing action in a given state. It tells, how was the action for achieving agent's goal.
d. **Policy (π)**. Policy is strategy agent uses to determine which action to be taken in each state. In Q-learning, policy is indirectly learned by maximizing action-values $\left(Q(s,a)\right)$.

2.2.2 Q-function and Q-learning equation

Action-value function $Q(s,a)$ estimates probable cumulative reward that can be obtained from state 's', taking action 'a', by following optimal policy thereafter. Value of $Q(s,a)$ is updated iteratively based on rewards and subsequent states agent experiences. Bellman equation defines relationship between Q-values and rewards.

$$Q(s,a) = E\left[r_{t+1} + \gamma \underset{a'}{max} Q(s_{t+1},a')\right] \tag{2.1}$$

Where, $Q(s,a)$ is action-value function for a given state-action pair and r_{t+1} is instant reward, agent receives after acting 'a' in state 's' and transitioning to next state s_{t+1}. γ ranges between 0 and 1 is discount factor, which determines how much future rewards are taken into account compared to immediate rewards. $\underset{a'}{max} Q(s_{t+1},a')$ represents maximum possible Q-value (best future expected reward) in next state s_{t+1}, considering all possible actions a'.

This equation is fundamental in Q-learning because it says that current Q-value is updated by considering immediate reward along with discounted future reward. Over time, as the agent explores environment, Q-values converge to optimal action-values, enabling the agent to learn the best possible policy.

2.2.3 Q-learning update rule

Q-learning updates action-value function $Q(s,a)$ based on agent's experiences is as follows.

$$Q(s_t,a_t) \leftarrow Q(s_t,a_t) + \alpha\left[r_{t+1} + \gamma \underset{a'}{max} Q(s_{t+1},a') - Q(s_t,a_t)\right] \tag{2.2}$$

Where, $Q(s_t,a_t)$ is current Q-value for state s_t and action a_t, r_{t+1} is reward achieved after performing action at in state s_t, leading to subsequent state s_{t+1}. $\underset{a'}{max} Q(s_{t+1},a')$ is maximum Q-value of next state s_{t+1}, representing best possible future return and α lying between 0 and 1 is learning rate, which regulates overriding of new information over old. γ is discount factor, indicating how much future rewards are taken into account relative to immediate rewards.

This update rule is based on temporal difference (TD) learning approach. State-action pair's Q-value is updated incrementally after each interaction with environment. This allows Q-learning to learn from each experience without needing to know the complete model of environment (transition and reward functions).

2.2.4 Exploration versus exploitation in Q-learning

Key challenges in Q-learning, and reinforcement learning in general, is balancing exploration i.e. trying new actions and exploitation i.e. selecting actions that the agent already knows will yield high rewards.

Exploration involves selecting actions that have not been tried before or that the agent is uncertain about. This can be done using a strategy such as epsilon-greedy.

In epsilon-greedy, the agent typically chooses the action having the highest Q-value (exploitation), but with probability ϵ, it selects random action (exploration). As the agent learns, ϵ is often decayed over time, allowing agent to shift from exploration to exploitation.

$$\text{Action Selection} = \begin{cases} \text{random action} & \text{with probability } \epsilon \\ \arg\max_a Q(s_t, a) & \text{with probability } 1 - \epsilon \end{cases}$$

Exploitation is focused on selecting action with highest Q-value in current state, ensuring the agent maximizes its expected reward.

2.3 CONVERGENCE OF Q-LEARNING

Q-learning is proven to results in optimal action-value if the following conditions hold.

- Agent explores environment sufficiently and updates each state-action pair infinitely often.
- Learning rate α decreases over time (e.g., α_t such that $\sum_t \alpha_t = \infty$ but $\sum_t \alpha_t^2 < \infty$)

Under these conditions, Q-learning will converge to optimal policy. However, in practice, Q-learning is often applied in environments where agent does not have infinite time to explore, and a sufficient approximation to optimal policy is considered acceptable.

2.4 DEEP Q - NETWORKS (DQN)

Deep Q - Networks (DQN) are a powerful algorithm in Reinforcement Learning (RL) that combine Q-Learning with deep NN to solve multifaceted tasks with multi-dimensional state spaces, such as video games and robotics [5, 6]. Q-Learning has been explained above and other aspects are as mentioned below.

a. **Deep Neural Networks.** Instead of using a table to store Q-values (which becomes infeasible for large state-action spaces), DQNs approximate Q-function using a deep neural network $Q(s,a;\theta)$, where θ represents network's parameters.

b. **Experience Replay**. To break correlation between consecutive experiences, DQNs store transitions (s,a,r,s') in replay buffer and sample minibatches while training. It improves stability and efficiency by reusing past experiences.

c. **Target Network**. DQNs maintain a separate target network $Q(s,a;\theta')$, that is regularly updated with weights of core network. This helps in stabilizing training by reducing oscillations in Q-value updates.

2.4.1 Algorithm

a. **Initialize**
   - Q-network $Q(s,a;\theta)$ through random values.
   - Target-network $Q(s,a;\theta')$ and set $\theta' = \theta$.
   - Replay buffer to store experiences.

b. **Repeat**
   - Initialize state s_0.
   - For every time step t.

   Select action a_t through ϵ-greedy policy.

$$a_t = \begin{cases} \text{random action} & \text{with probability } \epsilon \\ \arg\max_a Q(s_t,a;\theta) & \text{otherwise} \end{cases}$$

   Perform a_t, perceive reward r_t and subsequent state s_{t+1}.
   Store (s_t, a_t, r_t, s_{t+1}) in buffer
   Take a minibatch from buffer and update Q-network by minimizing loss.

$$L(\theta) = E_{(s,a,r,s')}\left[\left(r + \gamma \max_{a'} Q(s',a';\theta') - Q(s,a;\theta)\right)^2\right] \tag{2.3}$$

   Periodically update target network weights. $\theta' \leftarrow \theta$.

c. **End episode**

   In this method Q-value is updated using Bellman equation, like standard Q-learning, but with neural network's output. DQN has been effectively used in various areas, including video game (e.g., Atari game) and robotic control tasks.

2.5 POLICY GRADIENT METHODS

These optimize policy straightway [7]. They parameterize policy with a function approximator (e.g., neural network) and update it using gradients.

In this method, the aim is to optimize the likely cumulative reward, typically formulated as.

$$J(\theta) = E\left[\sum_{t=0}^{T} r_t\right] \tag{2.4}$$

Policy gradient is computed using likelihood ratio and reward-to-go.

$$\nabla J(\theta) = E[\nabla \log \pi\theta(a_t \mid s_t) G_t] \tag{2.5}$$

where G_t is cumulative reward starting from time t. Policy gradient methods are useful in multi-dimensional action spaces including complex environments, such as robotics and strategy games.

Policy Gradient Methods are a type of Reinforcement Learning algorithms where the agent learns a policy directly by optimizing a performance objective. Unlike value-based methods, which estimate a value function, policy-gradient methods parameterize and optimize policy itself. Key Concepts are as follows.

- **Policy.** A policy $\pi(a|s);\theta$ is a probability distribution of actions a, for given state 's', parameterized by θ. In deterministic policies, $\pi(s;\theta)$ maps states directly to actions.
- **Policy Optimization.** Aim is to optimize expected cumulative rewards.

  $$J(\theta) = E_{\tau\sim\pi\theta}\left[R(\tau)\right] \tag{2.6}$$

  where τ is a trajectory, generated through policy, and $R(\tau) = \sum_t \gamma^t r_t$ is discounted return.
- **Policy Gradient Theorem.** Gradient of $J(\theta)$ is as given below.

  $$\nabla_\theta J(\theta) = E_{\tau\sim\pi_\theta}\left[\nabla_\theta \log \pi_\theta(a \mid s) R(\tau)\right] \tag{2.7}$$

  This is derived using a likelihood ratio trick, enabling the gradient estimation without knowing the explicit form of $R(\tau)$.

2.5.1 Core algorithm: REINFORCE

a. **Initialize.**
   - Policy parameters θ.
   - Learning rate α.

b. **Repeat.**
   - Generate a trajectory τ = (s_0, a_0, r_0, s_1, a_1, r_1, ...) by following policy π_θ.
   - Compute total return $R_t = \sum_{k=t}^{T} \gamma^{k-t} r_k$ for each time step t.

c. **Update policy parameters using gradient.**

$$\theta \leftarrow \theta + \alpha\left[\nabla_\theta \log \pi_\theta(a_t \mid s_t) R(\tau)\right] \tag{2.8}$$

2.6 ACTOR-CRITIC METHODS

These make use of benefits of value & policy-based approaches. "Actor" updates policy, whereas "critic" examines policy through estimating value functions [8]. In this method, actor adjusts policy and evaluates how good the action taken was by approximating value function. In these methods, the actor updates policy using gradient, and these methods are versatile and can manage discrete and continuous action spaces. Core concepts in Actor-Critic Methods are as follows.

a. **Actor**. Represents policy $\pi(a|s;\theta_{actor})$, that determines probability of implementing an action 'a' in a state 's'. Actor is also responsible for improving policy by learning parameters θ_{actor} via policy gradient methods.
b. **Critic**. Estimates value function (e.g., $V(s;\theta_{critic})$ or $Q(s,a;\theta_{critic})$), which provides feedback to actor on how good the action was. Critic is also responsible for reducing variance in policy gradient by learning parameters θ_{critic}.
c. **Advantage Function**. Many actor-critic methods use advantage function.

   $$A(s,a) = Q(s,a) - V(s),$$

   This measures the performance of action '*a*' comparing with average action in state '*s*'.

2.6.1 Algorithm

a. **Initialization.**
   - Initialize parameters θ_{actor} for policy network.
   - Initialize parameters θ_{critic} for value function network.
b. **Repeat (for each episode).**
   - *Actor Step*. Use current policy $\pi(a|s;\theta_{actor})$ to generate actions and interact with environment.
   - *Critic Step*. Estimate value function $V(s;\theta_{critic})$ or $Q(s,a;\theta_{critic})$ using collected rewards.
   - *Policy Update*. Update actor using gradient of policy objective.

$$\nabla_{\theta_{actor}} J(\theta_{actor}) = E_{s\sim\pi,a\sim\pi}\left[\nabla_{\theta_{actor}} \log \pi(a \mid s;\theta_{actor}) A(s,a)\right] \tag{2.9}$$

- *Value Update*. Update critic by minimizing value function error.

$$L(\theta_{critic}) = E_{s,r,s'}\left[\left(r + \gamma V(s';\theta_{critic}) - V(s';\theta_{critic})\right)^2\right] \quad (2.10)$$

2.6.2 Key variants

a. **Advantage Actor-Critic** (A2C). It uses advantage function $A(s,a) = Q(s,a) - V(s)$ to improve actor's updates. A2C is a synchronous version, where multiple agents interact with their environments in parallel.
b. **Asynchronous Advantage Actor-Critic** (A3C). An asynchronous variant of A2C in which many agents interact with their individual environment independently. It also reduces correlation in updates and stabilizes learning.
c. **Deep Deterministic Policy Gradient** (DDPG). It is an extension of actor-critic methods for continuous action spaces. It is a deterministic actor. $\mu(s;\theta_{actor})$ and a critic. $Q(s,a;\theta_{critic})$. It relies on experience replay and a target network for stability.
d. **Soft Actor-Critic** (SAC). It incorporates an entropy term to encourage exploration and its objective becomes.

$$J(\pi) = E\left[\sum_t \gamma^t \left(r_t + \alpha H(\pi(.\mid s_t))\right)\right] \quad (2.11)$$

where H is entropy of policy and α controls the trade-off between exploration and exploitation.

2.7 MODEL-BASED METHODS

These involve constructing models of dynamics of environment, which encompasses transition probabilities and rewards [9]. They also allow the agent to simulate and plan subsequent actions, making it a powerful approach. It can either be deterministic or stochastic, capturing dynamics of environment and once a model is available, the agent can use it to simulate future trajectories, evaluate outcomes of different actions, and optimize its policy accordingly. Here the agent iteratively refines its model as it collects new experiences, aiming to improve the accuracy of its predictions and, consequently, its policy. The overview of Model-Based structure has been shown in Figure 2.3. Flow of a model-based RL algorithm follows the given steps below.

a. **Initialize**. Begin with initial policy and model of environment.
b. **Observe State**. Observe current state from environment.

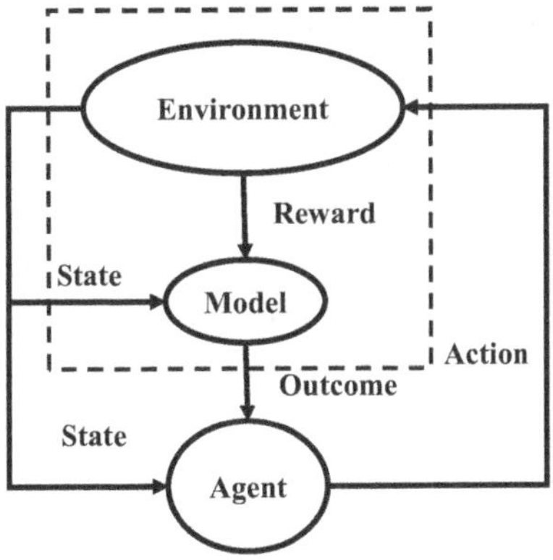

Figure 2.3 Overview of model-based structure.

c. **Update.** Based on new state transitions and rewards, refine model.
d. **Simulate.** Use model to simulate future states and rewards for different actions.
e. **Select Action.** Based on outcomes, choose action with largest expected reward.
f. **Observe Reward.** Execute action, observe outcome, and collect reward.
g. **Repeat.** Repeat cycle as new data is collected.

Model-based methods are further categorized into various approaches as follows.

2.7.1 Dynamic programming

These techniques assume complete knowledge about environment's model and can be used to compute optimal policies [10]. This method updates value functions iteratively until convergence, using Bellman equation to compute value of each state based on possible actions and their outcomes. Core Concepts of Dynamic Programming are as follows: -

a. **Markov Decision Process (MDP).** An MDP is defined by.
   - S (Set of states)
   - A (Set of actions)
   - P (s' | s, a) (Transition probability function)
   - R (s, a) (Reward function)
   - γ {Discount factor ($0 \le \gamma \le 1$)}
b. **Bellman Equations.** Foundation of DP methods is Bellman equations.
   - State-Value Function ($V(s)$).

$$V(s) = E_{a\sim\pi,s'\sim P}[(R(s,a) + \gamma V(s')] \tag{2.12}$$

- Action-Value Function ($Q(s,a)$).

$$Q(s,a) = R(s,a) + \gamma \sum_{s'} P(s' \mid s,a) \max_{a'} Q(s',a') \qquad (2.13)$$

c. **Policy**. A policy $\pi(a \mid s)$ maps states to actions and DP methods aim to find an optimal policy that maximizes expected cumulative rewards.

2.7.2 Key DP Algorithms in RL

a. **Policy Evaluation:** It computes value function $V^{\pi}(s)$ for policy π and iteratively updates $V(s)$ using Bellman expectation equation as follows.

$$V_{k+1}(s) = \sum_{a} \pi(a \mid s) \sum_{s'} P(s' \mid s,a)\left[R(s,a) + \gamma V^{\pi}(s)\right] \qquad (2.14)$$

b. **Policy Improvement:** It improves given policy π using value function $V^{\pi}(s)$ and generates a new policy as follows.

$$\pi'(s) = \arg\max_{a} \sum_{s'} P(s' \mid s,a)\left[R(s,a) + \gamma V^{\pi}(s)\right] \qquad (2.15)$$

c. **Policy Iteration:** It repeatedly alternates between evaluating and improving the policy until the policy converges to the optimal solution. The steps are as follows.
   - Start with initial policy π.
   - Evaluate $V^{\pi}(s)$using policy evaluation.
   - Improve π using policy improvement.
   - Repeat until π stabilizes.

d. **Value Iteration:** It merges policy evaluation and improvement by iteratively updating V(s) using Bellman optimality equation.

$$V_{k+1}(s) = \max_{a} \sum_{s'} P(s' \mid s,a)\left[R(s,a) + \gamma V_k(s')\right] \qquad (2.16)$$

Optimal policy is obtained from optimal value function.

$$\pi^*(s) = \arg\max_{a} \sum_{s'} P(s' \mid s,a)\left[R(s,a) + \gamma V^*(s')\right] \qquad (2.17)$$

Model Predictive Control strategy is used by the model for predicting future states and optimize control inputs over a finite horizon [11]. Here at each time step, the agent simulates several future trajectories based on the current state and evaluates expected outcomes. It then selects action that maximizes expected cumulative reward over horizon. MPC is widely used in robotics, autonomous vehicles, and industrial process control.

2.8 MONTE CARLO TREE SEARCH

MCTS is an algorithm which uses randomness and statistical sampling to make decisions in environments where the outcome of actions is uncertain or problem space is too large to explore exhaustively. It has gained prominence in Reinforcement Learning (RL), especially in domains like games and planning. It is effective in environments having large state spaces and branching factors [12]. Four Key Steps of MCTS, which are repeated iteratively are as follows.

a. **Selection**: Beginning from root node, the algorithm choses child nodes depending on selection policy until a leaf node is arrived. UCT balances exploration and exploitation.

$$UCT = Q(s.a) + C\sqrt{\frac{\ln N(s)}{N(s,a)}} \tag{2.18}$$

where $Q(s,a)$ is estimated value of action a from state 's', $N(s)$ is total number of visits to state 's', $N(s,a)$ is number of times action 'a' has been taken from state 's' and C is exploration constant.

b. **Expansion**: If chosen node is not terminal and has unvisited child nodes, one of them is added to the tree as a new leaf node.
c. **Simulation (Rollout)**: A simulation is executed from new leaf node to a terminal state using a default policy (e.g., random actions). Result of simulation is recorded as reward or payoff.
d. **Backpropagation**: Reward obtained from simulation is propagated back up tree, updating visit count and value estimates (Q) for each node and action along path.
e. **Final Decision**: After a specified number of iterations, action corresponding to most visited child of root node is chosen as best action.

2.8.1 Advantages of MCTS in RL

a. **Scalability**: MCTS does not need a complete model of environment and can work with simulations, making it suitable for large state-action spaces.
b. **Anytime Algorithm**: MCTS can return a result even if interrupted before completion, and more computation improves solution.

c. **Balancing Exploration and Exploitation**: UCT formula ensures that both well-explored and less-explored branches are considered.
d. **Adaptability**: MCTS can handle stochastic environments and domains with imperfect information.

2.8.2 Learning models from data

In many real-world applications, the environment's dynamics are not fully known. Model-based methods can be enhanced by learning model from data [13]. In this method, agent collects data from its interactions and uses regression techniques or neural networks to approximate transition dynamics including reward functions and after model is trained, the same can be used with planning algorithms like value iteration or MPC to make informed decisions.

2.9 CHALLENGES IN MODEL-BASED METHODS

Following are some of the challenges faced in Model based methods. -

a. **Model Accuracy**. Performance of model-based methods severely depends on accuracy of trained model.
b. **Computational Complexity**. Algorithms are computationally intensive, particularly in high-dimensional state spaces.
c. **Exploration**. Model-based methods need effective exploration strategies to gather informative data for model learning, balancing exploration, and exploitation.

2.9.1 Applications of model-based methods

Below are some of the applications pertaining to Model based methods:

a. **Robotics**. For trajectory planning and control, robots can learn and adapt to dynamic environments.
b. **Autonomous Vehicles**. In navigation and decision-making processes, safe and efficient driving can be enabled.
c. **Healthcare**. Personalized treatment planning, where models can predict patient responses to different interventions.
d. **Game AI**. In strategic games, allowing agents to simulate and evaluate potential moves before making decisions.

2.10 EXPLORATION VS EXPLOITATION

Key challenge in RL is balancing exploration and exploitation. Following are suggested strategies to manage this trade-off.

2.10.1 Epsilon-Greedy

This is a fundamental approach used in RL to balance exploration-exploitation trade-off. Strategy is particularly effective in environments where an agent needs to learn best actions for taking in different states through trial and error [14]. Following paragraphs explore mechanics of epsilon-greedy strategy, its implementation, and its advantages and disadvantages.

In reinforcement learning, agent has to decide whether to make use of its current knowledge (choosing action known to yield highest reward) or explore new actions that might yield greater rewards. Epsilon-greedy strategy provides a simple method for incorporating exploration into the decision-making process. Epsilon-greedy strategy operates as follows:

a. **Set Epsilon (ϵ)**: Epsilon is a small value between 0 and 1 that represents probability of exploration. A common choice might be $\epsilon = 0.1$, meaning there is a 10% chance of exploring new actions.
b. **Action Selection**: At each decision point, the agent selects the action that it believes has the highest expected reward i.e. exploitation with $1-\epsilon$ probability and the agent randomly selects an action from action space i.e. exploration with probability ϵ. This approach ensures that the agent occasionally tries out less-known actions, leading to discovering improved strategies or rewards. Following are implementation Steps
    - Start with an initial estimate for Q-values of all actions (often zeros).
    - Iterate through episodes.
    - For each episode, observe current state 's'.
    - Apply epsilon-greedy strategy to choose an action 'a'.
    - Generate a random number between 0 & 1.
    - Random number, if less than 'ϵ', choose a random action; otherwise, select action with largest Q-value.
    - Execute chosen action, check reward 'r' along with new state 's′'.
    - Update Q-value for action taken utilizing received reward including new state.
    - Repeat.
    - Continue this process until termination condition is achieved.
c. **Advantages**: Epsilon-greedy strategy is easy to implement and understand, making it a choice for introductory RL problems and it effectively balances exploration and exploitation, allowing agents to discover new strategies while still capitalizing on known rewards.
d. **Disadvantages**: Static value of 'ϵ' can lead to suboptimal exploration over time. As learning progresses, the agent may continue to explore even when it has enough knowledge, or it may become overly exploitative if 'ϵ' is too low. Secondly, randomness in action selection can lead to inefficient learning, especially in environments where best actions are close together in terms of Q-value.

e. **Variants:** To address some of disadvantages of epsilon-greedy strategy, several variants have been proposed, which are as follows:
   - **Decay Epsilon.** Gradually reduce 'ϵ' over time (e.g., from 1.0 to 0.1) to encourage exploration early on and shift toward exploitation as learning progresses.
   - **Adaptive Epsilon.** Adjust 'ϵ' dynamically based on agent's performance, allowing it to explore more when it is uncertain and exploit more when it is confident.
   - **Softmax Action Selection:** Instead of a fixed probability, this method choses actions based on their assessed values, assigning higher probabilities to actions having higher Q-values while allowing for exploration.

2.11 SOFTMAX ACTION SELECTION

This is an advanced technique used in RL to balance exploration-exploitation trade-off more dynamically than simpler methods like epsilon-greedy [15]. Instead of selecting actions randomly or purely based on a fixed probability, Softmax assigns probabilities to actions depending on their calculated values.

In Softmax action selection method, each action's selection probability is influenced by its estimated value (e.g., Q-value) relative to other actions. This guarantees that actions with larger estimated rewards are likely to be chosen, while still maintaining some probability of selecting lower-value actions to facilitate exploration. Softmax selection process can be described in following steps:

a. **Calculate Action Preferences:** For a given state 's', compute estimated values (Q-values) for all available actions a (denoted as $Q(s,a)$).
b. **Compute Softmax Probabilities:** Use Softmax function to transform these values into probabilities. formula for Softmax probability $P(a \mid s)$ of action 'a' is given by.

$$P(a \mid s) = \frac{e^{\frac{Q(s,a)}{\tau}}}{\sum_b e^{\frac{Q(s,b)}{\tau}}} \tag{2.19}$$

   where $Q(s,a)$ is Q-value for action 'a' in state 's'. 'τ' controls level of exploration. A higher 'τ" leads to more uniform probabilities (more exploration), while a lower 'τ" makes selection greedier.
c. **Select Action.** Sample an action 'a' according to calculated probabilities.

2.11.1 Advantages

Softmax balances exploration and exploitation based on assessed action values, allowing for a more informed action selection. As Q-values are updated, action probabilities also adjust continuously, allowing the agent to adjust to changes in environment and unlike epsilon-greedy, which has abrupt shifts between exploration and exploitation, Softmax provides a smoother gradient of action selection probabilities.

2.11.2 Disadvantages

Performance of Softmax action selection is sensitive to choice of temperature parameter 'τ'. Selecting an inappropriate value can lead to either excessive exploration or premature convergence to suboptimal policies. Computing Softmax probabilities requires exponentiation and normalization, which can introduce computational overhead, especially in environments having large number of actions.

2.11.3 Variants and extensions

Several variants and extensions of Softmax action selection method may be employed for enhancing performance. These are mentioned below.

a. **Adaptive Temperature**. Instead of using a fixed 'τ", it can be adjusted dynamically based on agent's learning progress. For example, 'τ" can be decreased as agent gains confidence in its Q-value estimates.
b. **Boltzmann Exploration**. This is a more general form of Softmax that uses a temperature parameter to modulate exploration-exploitation balance further, allowing for more flexibility in action selection.
c. **Action Value Networks**. In deep RL, neural networks may be used to estimate Q-values, and Softmax selection can be applied to the output of these networks for action selection.

2.11.4 Applications

Softmax action selection is widely used in environments where agents need to learn complex strategies (e.g., board games, video games), Softmax helps balance the exploration of latest strategies with the exploitation of known successful actions. Also in robot control tasks, Softmax can assist in selecting actions that optimize performance while exploring new movements or behaviors.

2.12 UPPER CONFIDENCE BOUND (UCB)

This method is used in RL to balance exploration and exploitation. UCB approaches decision-making by taking into account the estimated value of action along with uncertainty associated with that estimate. This strategy allows agents to make informed decisions while ensuring they explore less-tried options [16].

The core idea behind UCB is to choose actions that maximize expected rewards including accounting for uncertainty in estimates of those rewards. By maintaining the confidence interval around estimated value of every action, UCB encourages exploration of actions that have high uncertainty. UCB algorithm can be outlined in following steps.

a. **Initialize Values.** For each action a, initialize estimated reward $Q(a)$ and count of times action has been chosen $N(a)$.
b. **Action Selection.** At every time step 't', select action a_t based on following formula.

$$a_t = \arg \max_a \left(Q(a) + c\sqrt{\frac{\ln t}{N(a)}} \right) \tag{2.20}$$

where, $Q(a)$ is calculated value of action '*a*'. $N(a)$ is number when action '*a*' has been chosen. '*c*' is constant which determines level of exploration; larger values of '*c*' encourage more exploration.
c. **Update Estimates.** After acting a_t, check reward r_t and update estimated value and count.

$$\mathrm{N}(\mathrm{a}_t) \leftarrow \mathrm{N}(\mathrm{a}_t) + 1 \tag{2.21}$$

$$\mathrm{Q}(\mathrm{a}_t) \leftarrow \mathrm{Q}(\mathrm{a}_t) + [1/\mathrm{N}(\mathrm{a}_t)](\mathrm{r}_t - \mathrm{Q}(\mathrm{a}_t)) \tag{2.22}$$

d. **Repeat.**
Continue process for predetermined number of iterations or till convergence.

2.12.1 Advantages

UCB algorithms come with strong theoretical performance guarantees, showing that they achieve logarithmic regret over time, meaning they learn to select optimal actions efficiently. UCB naturally balances need to discover actions with more uncertainty and exploit actions with identified high rewards, promoting a more strategic exploration. Also, UCB approach is straightforward to implement and may be applied in numerous settings, making it a popular choice for several RL problems.

2.12.2 Disadvantages

Performance of UCB can be sensitive to the choice of exploration constant 'c'. A poorly chosen 'c' may lead to excessive exploration or premature convergence. UCB algorithms typically assume that environment is stationary, meaning reward distributions do not change over time. This assumption may not hold in dynamic environments. Also, in environments having a large number of actions, calculating UCB for each action may become computationally expensive.

2.12.3 Variants and extensions

Several variants of UCB algorithm exist to improve its performance or adapt it to specific scenarios. Some of these are mentioned below. -

a. **UCB1-Tuned.** This adjusts exploration term depending on empirical adjustment of rewards, providing additional superior exploration strategy [17].
b. **Bayesian UCB.** This integrates Bayesian inference for modeling uncertainty accurately, leading towards improved action selection in some contexts [18].
c. **Contextual UCB.** Extending UCB to contextual bandits, where agent has access to additional context or features, allows for more informed action selection based on current state [19, 20].

2.13 APPLICATION OF RL IN DECISION MAKING

We have seen in previous sections that in reinforcement learning, decisions are made based on an agent's interactions with environment, where the agent selects actions with the goal of maximizing cumulative rewards. This decision-making process involves policy, which guides agent on which action to take in each state. The agent explores environment, takes actions, observes resulting state changes, and receives rewards that indicate quality of those actions. Over time, the agent uses algorithms, such as Q-learning or policy gradients, to update its policy and improve its decision-making by learning which actions yield the highest rewards in specific situations. This repetitive process permits the agent to adapt and optimize its decisions to achieve long-term success, often balancing exploration of actions with exploitation of rewarding actions.

Therefore, in RL, decision making is the process where the agent selects actions to optimize cumulative rewards by learning from environment through trial and error.

Consider a simple example of a game where a robot is placed in a grid, and its goal is to reach "exit (E)". In Figure 2.4, S indicates start position of robot, E indicate Exit i.e. goal position and W indicates Wall i.e. obstacle.

S		W		
	W		W	
		W		E
			W	

Figure 2.4 Robot is placed in a Grid at cell 'S'.

Flow of Decision-Making using RL starts with initializing i.e. start at 'S' with no prior knowledge then robot identifies its current position on grid (Observe State). This is followed by selecting an action (e.g., move up, down, left, right) randomly by robot (Select action) and then observer reward i.e. positive when moving closer to exit, negative when hitting a boundary wall and highly positive reward while reaching at exit. Based on rewards, policy is updated. This is repeated till shortest path to exit is discovered.

In a nutshell, the robot starts by exploring random moves, receiving feedback as rewards for each step. It then uses this feedback to update its strategy. Over time, the robot learns a "policy," which is a set of rules that tells it the best action to be taken in each cell. This policy is updated iteratively and each time the robot gets feedback from environment, it adjusts its understanding of which actions are most beneficial. As it gains more experience, the robot learns to avoid unnecessary moves, taking only actions that maximize its rewards and reach the exit more efficiently. Through this learning process, the robot achieves optimal decision-making by continually refining its choices based on the cumulative feedback it receives.

2.14 AN EXAMPLE OF DECISION MAKING

Reinforcement Learning has applications in many fields, including gaming, robotics, finance, and energy etc. The following is an example of a simple reinforcement learning model using MATLAB to make decisions in a basic environment. In this example, Q-learning algorithm has been implemented to enable the agent to learn optimal policy for maximizing rewards over time. In this example, simple grid environment has been simulated and the agent's reward has been plotted over time to show learning process in order to maximize reward by reaching a goal. RL algorithm Q-learning has been used and the results of the learning progress has been shown in following diagrams.

Figure 2.5 shows cumulative reward that the agent received in each episode. A positive trend indicates that the agent is learning to reach goal more efficiently, as it gathers higher rewards over time.

Figure 2.6 shows a moving average of times the agent reached goal across episodes. A higher average over time indicates the agent is becoming more successful in reaching goal.

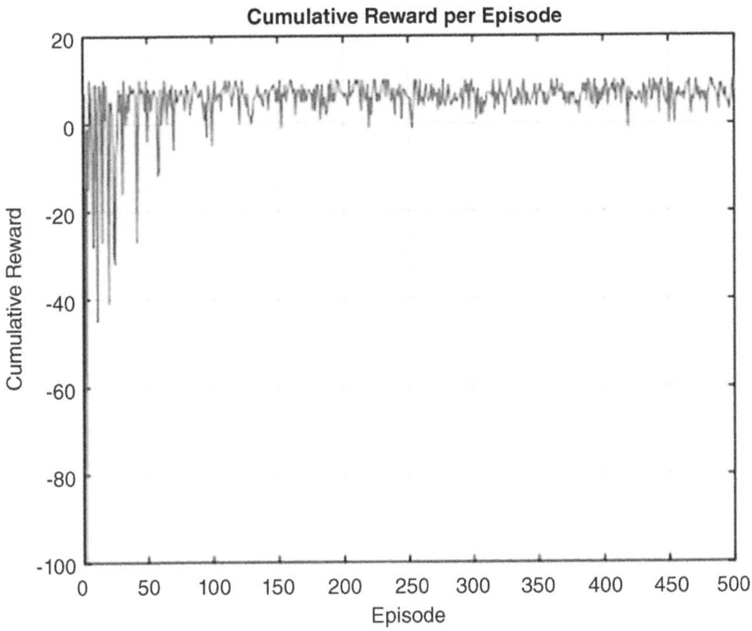

Figure 2.5 Cumulative reward that the agent received in each episode.

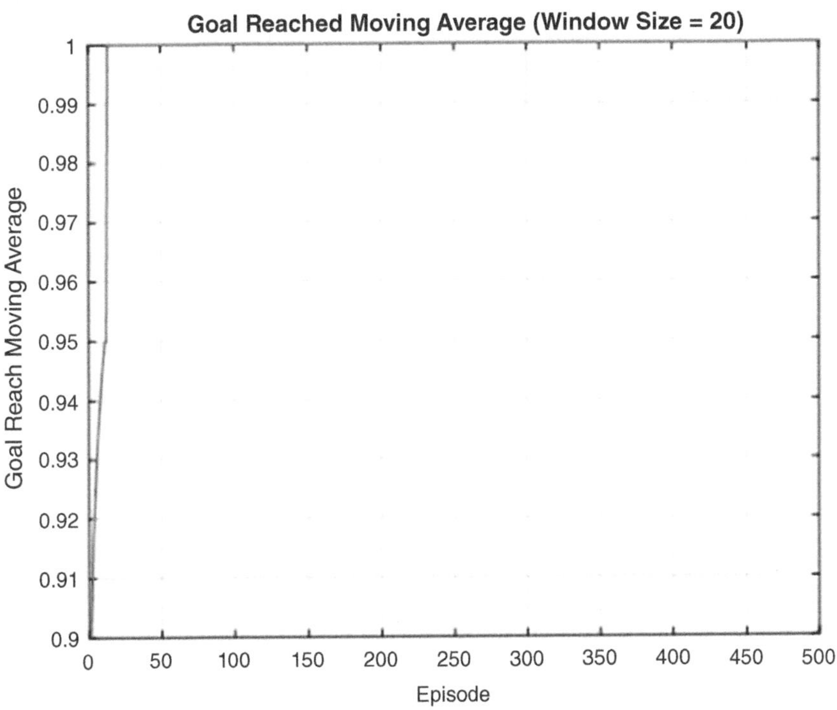

Figure 2.6 Moving average of times the agent reached goal.

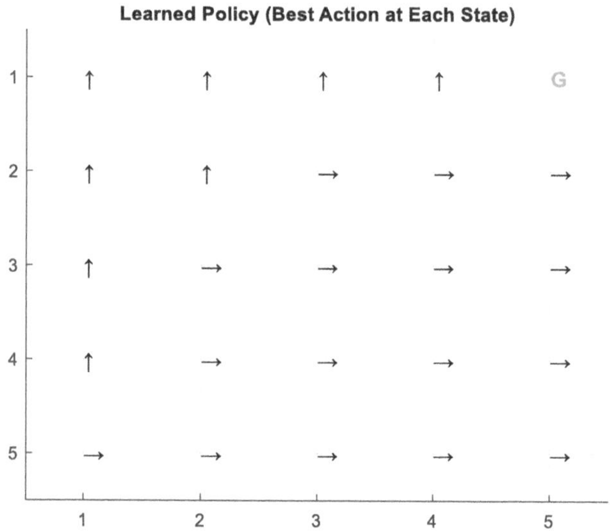

Figure 2.7 Arrows showing the best action agent has learned for each state.

Figure 2.7 plots grid displaying arrows showing the best action agent has learned for each state, based on Q-values. Position labelled 'G' is goal state. Arrows represent optimal path learned by the agent. This example demonstrates the reinforcement learning process in which the agent makes better decisions in an uncertain environment. Through training, the agent learns optimal policy, which is represented here by Q-values and visualized as arrows indicating best actions.

REFERENCES

[1] Hammoudeh, Ahmad. (2018). A concise introduction to reinforcement learning. https://doi.org/10.13140/RG.2.2.31027.53285

[2] Lu, Fengyi, Zhou, Guanghui, Chao, Zhang, Liu, Yang & Taisch, Marco (2025). Integrated optimization of multi-pass cutting parameters and tool path with hierarchical reinforcement learning towards green manufacturing. *Robotics and Computer-Integrated Manufacturing*. 91. 102824. https://doi.org/10.1016/j.rcim.2024.102824

[3] T. Degris, P. M. Pilarski and R. S. Sutton, Model-Free reinforcement learning with continuous action in practice. *2012 American Control Conference (ACC)* (pp. 2177–2182). 2012. https://doi.org/10.1109/ACC.2012.6315022

[4] Jang, Beakcheol, Kim, Myeonghwi, Harerimana, Gaspard, & Kim, Jong (2019). Q-learning algorithms. A comprehensive classification and applications. *IEEE Access*, 1–1. https://doi.org/10.1109/ACCESS.2019.2941229

[5] M. A. Samsuden, N. M. Diah and N. A. Rahman, A review paper on implementing reinforcement learning technique in optimizing games performance. *2019 IEEE 9th International Conference on System Engineering and Technology*

(ICSET), Shah Alam, Malaysia, 2019, pp. 258–263. https://doi.org/10.1109/ICSEngT.2019.8906400
[6] Sewak, Mohit. (2019). *Deep Q network (DQN), Double DQN, and dueling DQN. A step towards general artificial intelligence.* https://doi.org/10.1007/978-981-13-8285-7_8
[7] Peters, Jan. (2010). Policy gradient methods. *Scholarpedia*, 5, 3698. https://doi.org/10.4249/scholarpedia.3698
[8] Zhongyang, Ming, Huaguang, Zhang, Juan, Zhang, & Xiangpeng, Xie (2023). A novel actor–critic–identifier architecture for nonlinear multiagent systems with gradient descent method. *Automatica*, 155, 111128. https://doi.org/10.1016/j.automatica.2023.111128
[9] Luo, Fan-Ming, Xu, Tian, Lai, Hang, Chen, Xiong-Hui, Zhang, Weinan, & Yu, Yang. (2024). A survey on model-based reinforcement learning. *Science China Information Sciences*, 67. https://doi.org/10.1007/s11432-022-3696-5
[10] Liu, D., Ha, M., & Xue, S. (2022). State of art of adaptive dynamic programming and reinforcement learning. *CAAI Artificial Intelligence Research*, 1(2), 93–110. https://doi.org/10.26599/AIR.2022.9150007
[11] Morcego, Bernardo, Yin, Wenjie, Boersma, Sjoerd, van Henten, Eldert, Puig, Vicenç, & Sun, Congcong (2023). Reinforcement learning versus model predictive control on greenhouse climate control. *Computers and Electronics in Agriculture*, 215, 108372. https://doi.org/10.1016/j.compag.2023.108372
[12] James, Steven, Konidaris, George & Rosman, Benjamin (2017). An analysis of Monte Carlo tree search.
[13] Fruit, Ronan. (2019). *Exploration-exploitation dilemma in Reinforcement Learning under various form of prior knowledge.* Diss. Université de Lille 1, Sciences et Technologies; CRIStAL UMR 9189.
[14] Tokic, Michel. (2010). Adaptive ε-greedy exploration in reinforcement learning based on value differences. 203–210. https://doi.org/10.1007/978-3-642-16111-7_23
[15] Vamplew, Peter & Dazeley, Richard & Foale, Cameron (2017). Softmax exploration strategies for multiobjective reinforcement learning. *Neurocomputing*, 263. https://doi.org/10.1016/j.neucom.2016.09.141
[16] Kovari, Balint, Pelenczei, Bálint, & Bécsi, Tamás (2023). Enhanced experience prioritization. A novel upper confidence bound approach. *IEEE Access*, 1–1. https://doi.org/10.1109/ACCESS.2023.3339248
[17] Qu, Chang. (2024). Enhancing UCB-tuned and asymptotically optimal UCB algorithms through weighted average techniques in multi-armed bandit scenarios. Highlights in science, *Engineering and Technology*, 94. 187–194. https://doi.org/10.54097/8mb1rb15
[18] Kang, Pyungwon, Tobler, Philippe N., & Dayan, Peter (2024). Bayesian reinforcement learning. A basic overview, *Neurobiology of Learning and Memory*, 211, 107924. https://doi.org/10.1016/j.nlm.2024.107924
[19] Foster, Dylan, & Rakhlin, Alexander. (2020). *Beyond UCB. Optimal and Efficient Contextual Bandits with Regression Oracles.*
[20] Sivamayil, K.; Rajasekar, E.; Aljafari, B.; Nikolovski, S.; Vairavasundaram, S., & Vairavasundaram, I. A Systematic Study on Reinforcement Learning Based Applications. *Energies* 2023, *16*, 1512. https://doi.org/10.3390/en16031512

Chapter 3

Metamaterial-integrated pattern reconfigurable antennas

Reena Pant

3.1 INTRODUCTION

Antenna is one of the vital components without which wireless communication cannot be possible. It acts as an interface between the radio system and the external environment, facilitating the radiation or capture radio signals. In today's era of communication, there is a growing need to integrate multiple antennas within a single structure to accommodate different wireless services to save installation space and reduce fabrication cost and must be fast, efficient, and capable of shifting rapidly between different modes as per the requirements.

The ability of an antenna to dynamically and controllably adjust its operating frequency, radiation patterns, and/or polarization called as reconfigurable antenna, which classify them as:

a. Frequency reconfigurable antenna,
b. Pattern reconfigurable antenna,
c. Polarization reconfigurable antenna,
d. Compound reconfigurable antennas, which combine more than one type of reconfiguration for greater adaptability.

Pattern Reconfigurable Antennas (PRAs): The ability of an antenna to dynamically change its radiation pattern or direction is called pattern reconfigurability. Such antenna can focus or steer their signal in specific directions or cover particular areas, making it highly versatile and adaptable to various operational needs. In crowded environments, the high density of users leads to limited bandwidth and increased interference. Similarly, in automotive applications, where the transmitting or receiving antenna is in motion, maintaining a stable connection becomes challenging. In these scenarios, pattern reconfiguration allows antennas to direct signals toward specific users or areas, thereby improving signal coverage, reducing interference, and maintaining reliable communication links. This significantly enhances overall system efficiency, adaptability, and performance in complex and dynamic environments.

DOI: 10.1201/9781003568995-3

3.1.1 Classification of PRAs

PRAs can be categorized into beam-steering (beam-scanning) and beam-shaping (beam-forming). Here, beam-steering the ability to change the antenna's radiation direction of the primary lobe and beam-shaping involves altering the shape of the radiation pattern to meet system requirements. It can also be categorized, based on their mechanisms and design approaches, for altering radiation patterns. Reconfiguration by physically replacing or reorienting different components of the antenna is called mechanical reconfiguration.

Phased array PRAs use phase-shifters or time delays at each element in an array to electronically steer the beam in different directions which provides precise beam control and fine-tuning without any physical movement but often suffers significant losses, high complexity, and high cost.

Electronic reconfiguration is achieved using tunable devices like PIN diodes, MEMS switches, varactors etc. Factors such as fast switching, high efficiency, low cost, and ease of integration are critical when selecting a tuning device for pattern reconfiguration.

Metamaterial-based antennas have garnered significant attention due to their ability to direct RF currents in a controlled manner over the antenna surface, enabling the modification of conventional antenna properties to achieve specific design goals. Various tuning methods dynamically adjust the performance of metamaterial-based antennas by altering the capacitance or conductance of unit cells, providing additional degrees of freedom and enabling the development of adaptive wireless communication systems.

PRAs are valuable in wireless communication networks, where they reduce interference and improve signal quality by directing energy toward desired locations. In satellite communications, PRAs optimize data transmission by dynamically adjusting to changing user positions or atmospheric conditions. They are also critical in military applications, enabling secure, focused communication links and countering jamming or interception attempts. Their adaptability significantly improves system performance, reliability, and security in increasingly complex communication environments. Reconfiguration also addresses multipath fading, a major challenge in urban communication systems [1–2].

This chapter highlights recent advancements in metamaterial-based PRAs, focusing on how metamaterial structures enhance the adaptability and performance of antenna systems in dynamic environments. The review explores various aspects, including types of pattern reconfigurability, design architectures utilizing metamaterials, and enabling mechanisms such as tunable materials and phase-change elements. Research published over the past two decades emphasizes key performance parameters—such as gain, directivity, and radiation pattern control—critical for wireless communication applications. Technical challenges and future opportunities in integrating metamaterials for advanced pattern reconfiguration are also addressed.

The primary objective of this chapter is to provide a comprehensive analysis of metamaterial-based PRAs, examining how metamaterial structures contribute to improved beam-steering, radiation pattern control, and overall antenna performance. By reviewing recent design innovations, underlying reconfigurability mechanisms, and application-specific advantages, the chapter aims to consolidate current research findings, identify key performance metrics, and discuss the challenges and future potential of metamaterials in advancing reconfigurable antenna technologies.

3.2 METAMATERIALS

All materials found in nature have positive permittivity, permeability and a refractive index as shown in Figure 3.4. In metamaterial, 'meta' means beyond, meaning a material that shows certain properties that are not found in materials available in nature. The concept of metamaterial was first introduced by Mandelstam and Veselage [3–4] in 1945 and the experimental proof of negative permeability index was first presented by Smith and Pendry [5] 1999 by exotic structures. It is an engineered composite material which has properties like negative permittivity, permeability and negative refractive index. It is also called as negative index material or double negative media or left-handed material or backward wave media. Metamaterials are composed of very small unit cells arranged periodically where each cell behaves like a resonator that interacts with an external electromagnetic wave. In any engineered material, the properties of the material such as blocking, absorbing, enhancing or bending waves can be altered by changing the size, geometry and arrangement of its unit cells. This allows electromagnetic waves or acoustic waves to be easily manipulated within the material. The size, geometry, and arrangement of the cell is responsible for manipulating electromagnetic and acoustic waves by blocking, absorbing, enhancing, or bending waves to achieve desired properties that is not possible with conventional materials. Because size and spacing of each cell is much smaller than the wavelength of incoming electromagnetic radiation it is able to minimize the size of electronic components.

3.2.1 Classification of Metamaterials

Metamaterials can broadly classify based on their electromagnetic response as:

i. Electromagnetic metamaterials (EMMs): are classified based on their electromagnetic response, especially on permittivity and permeability variation as shown in Figure 3.1.
   a. Epsilon negative (ENG): stand for materials which have a negative permittivity but positive permeability and support only certain types of wave propagation. ENG exhibit unusual properties like

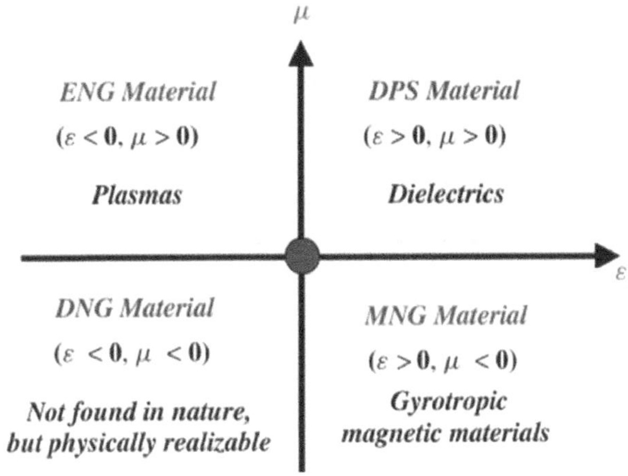

Figure 3.1 Classification of materials on the basis of its permittivity and permeability.

negative refraction or reversed Doppler effects. It allows strong coupling with electromagnetic waves, enabling surface plasmon polaritons at interfaces between ENG materials and positive-permittivity materials. Electromagnetic waves are evanescent within ENG materials because the wave vector becomes imaginary, leading to no propagation of wave inside. An array of thin conductors arranged in a specific manner can produce negative permittivity as shown in Figure 3.2. ENG loaded electrically small monopole antenna is shown in Figure 3.3.

b. Mu Negative (MNG): stand for materials which have a negative permeability but positive permittivity. This negative permeability occurs in specific frequency ranges due to resonance like the split-ring resonator (SRR) structure. Electromagnetic waves do not propagate

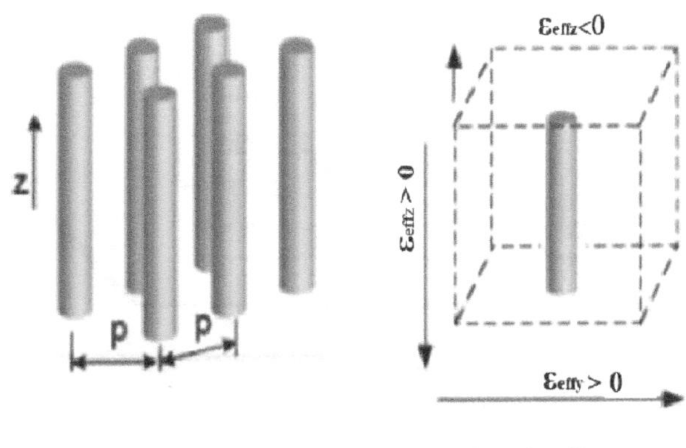

Figure 3.2 An array of thin conductor and unit cell with effective permittivity.

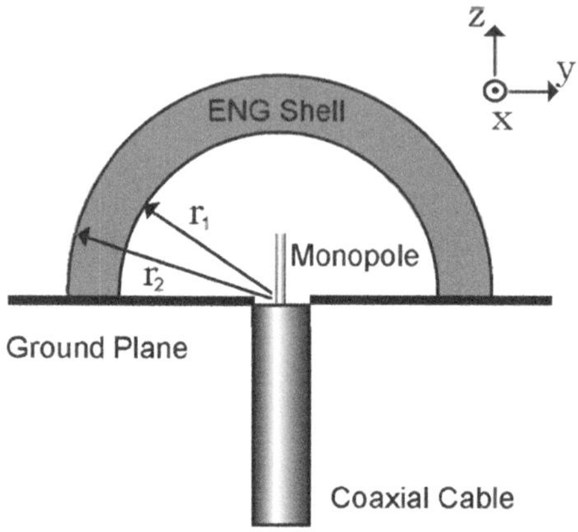

Figure 3.3 Electrically small antenna design in the structure of monopole and ENG environment.

through the material when only permeability is negative; instead, they decay exponentially, producing evanescent fields. In materials with negative permittivity like MNGs, the SRR is the most commonly used structure. An SRR unit cell consists of two concentric metallic rings, which can be either circular or square, separated by a small gap. Each ring features a narrow slot positioned 180 degrees apart on opposite sides. The gap between inner and outer rings functions as a capacitor, while the rings themselves serve as inductors. Together, they form an LC resonant circuit. Since the SRR has lower wavelength than its size due to its quasi-static resonance nature. It may be utilized to create tiny antennas. An example of printed antenna with a substrate from MNG is shown in Figure 3.4.

c. Double negative (DNG): stands for materials where both electric permittivity and magnetic permeability are negative over a specific

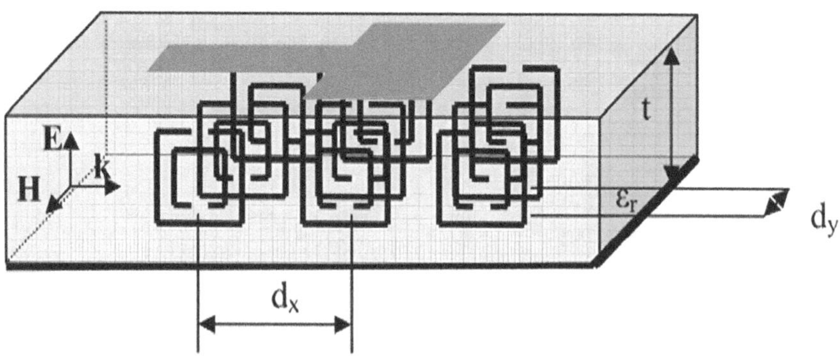

Figure 3.4 The printed antenna with a substrate form MNG.

frequency range. The waves bend opposite to the normal when entering the material, exhibit negative refraction.

d. Near zero index (NZI): in such type of materials electromagnetic waves travel with an extremely large wavelength inside materials, effectively creating a quasi-static field which leads to near uniform phase distribution within the material.
e. Epsilon near zero (ENZ): materials have near to zero permittivity, leading to near infinite wavelength within the material. This property enables electromagnetic waves to propagate minimal phase variation, allowing for effective wave confinement and redirection. This makes it easier to dynamically steer the antenna's radiation pattern or focus energy in desired directions.
f. Magnetic near zero index (MNZI) is an engineered material whose magnetic permeability and/or refractive index are close to zero at specific frequencies. When permeability of such material approaches almost zero, the magnetic response of the material becomes extremely weak, and electromagnetic waves can propagate through it with very little interaction with the magnetic field, causing electromagnetic waves to experience minimal phase variation while traveling through the material. These materials exhibit unique electromagnetic properties, enabling unprecedented control over wave propagation, impedance matching, and radiation characteristics.

ii. Resonant metamaterials: are classified based on their electromagnetic response, primarily determined by their resonant structures. These resonances significantly affect how they interact with electromagnetic waves. Here is a classification based on their response:
   a. Electric Resonant Metamaterials: Thin wires or other geometries that induce strong electric field localization exhibit electric dipole resonance when the structure provides negative permittivity at specific frequencies. Such structures are used to design antennas, waveguides, and enhance electric field interactions. Figure 3.5 (a) shows resonant electrical metamaterials.
   b. Magnetic Resonance Metamaterials: Split-ring resonators (SRRs) and labyrinth resonators exhibit magnetic dipole resonance when structure provides negative permeability at specific frequencies. Such structures are used to manipulate magnetic fields and in cloaking applications. Figure 3.5(b) shows resonant magnetic metamaterials.
   c. Electromagnetic (Dual) Resonant Metamaterials: Combinations of thin wires and SRRs exhibit simultaneous electric and magnetic resonance when structure provides a negative refractive index at specific frequencies. Such structures are used to manipulate magnetic fields, enable cloaking, and negative-index applications.

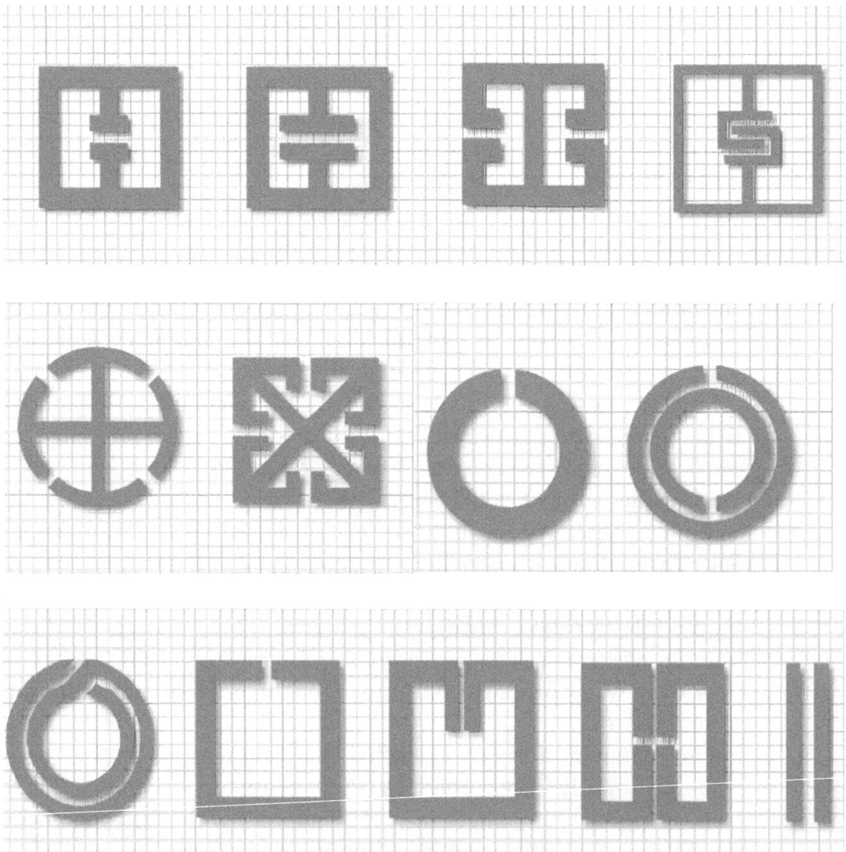

Figure 3.5 Examples of different patch shapes for (a) resonant electrical metamaterials and (b) magnetic metamaterials.

iii. Metasurfaces: this category includes engineered materials that are two-dimensional and extremely thin.
   a. FSS: structures are planar periodic arrays of conductive elements on a dielectric substrate that allow or block specific frequencies, acting as spatial filters.
   b. High impedance surface (HIS): Are specialized materials designed to control electromagnetic waves by controlling specific impedance, typically approaching infinity, which prevents wave propagation in certain directions while allowing it to pass others.
   c. AMC is an engineered material where it exhibits properties which are very close to perfect magnetic conductor (PMC). The AMC introduces almost a zero-degree phase shift.

d. Metasurfaces: are two-dimensional thin layer engineered materials designed to manipulate electromagnetic waves using subwavelength-scale structures. These materials can control electromagnetic waves at the nano-scale by altering the phase of incoming waves, adjusting wave intensity, and converting between linear and circular polarization. They can also deflect electromagnetic waves in unusual directions, defying Snell's law of reflection and refraction by introducing abrupt phase shifts.

iv. Nanomaterial based metamaterials: by integrating the special properties of nano-materials with metamaterial designs, these structures can manipulate electromagnetic waves in novel ways. These materials can exhibit negative permittivity and/or permeability in specific frequency ranges, leading to negative refractive index.

a. Liquid crystal metamaterials: combine the unique properties of liquid crystals such as anisotropy, tunability, and fluidity with the engineered structure of metamaterials. These hybrid systems enable dynamic control of electromagnetic waves and are used in applications requiring tunable or reconfigurable functionality.

b. Graphene metamaterials: when graphene's unique behavior as a two dimensional material with high conductivity, tunability, and flexibility combine with the engineered structure of metamaterials it becomes an excellent candidate for metamaterial applications across a wide frequency spectrum, especially in terahertz, infrared, and optical ranges.

3.2.2 Metamaterial based PRAs

Radiating structures and feeding networks are the two primary building blocks for designing reconfigurable antennas. Reconfiguration can be achieved either by making the feeding network itself reconfigurable or by employing other techniques to modify the antenna's radiation properties. Pattern reconfigurability can be classified into two categories: reconfigurable feeding networks and non-reconfigurable feeding networks. The most common approach for designing a reconfigurable feeding network involves the use of phased-array antennas. Other designs incorporate various switching mechanisms, such as PIN diodes or MEMS switches, into the feeding network to achieve reconfigurability. In the case of non-reconfigurable feeding networks, a fixed radiating source is combined with structures where reconfiguration is introduced by modifying the radiating element. This is achieved by integrating electrical switches like PIN diodes or varactor diodes into the patch, metamaterials, or FSSs.

Metamaterial structures offer diverse options for designing pattern-reconfigurable antennas, providing unique advantages such as enhanced reconfigurability, control over wave behavior, and ease of integration. Depending on how they interact with incident electromagnetic waves, metamaterials are categorized into electromagnetic metamaterials, resonant metamaterials, metasurfaces (including HISs, FSSs, and AMCs), and nanomaterial-based metamaterials like liquid crystal and graphene-based metamaterials.

Traditionally, metamaterials used for pattern reconfigurability required mechanical systems, such as motors, which made them bulky and less practical. To overcome this challenge, intelligent metasurfaces (IMS) have been widely adopted. IMS systems enable electrical reconfiguration using components like PIN diodes, microfluidic systems, MEMS switches, or FETs, resulting in compact and versatile designs [6]. Metamaterials that exhibit a negative refractive index characterized by negative permittivity and negative permeability are designed using two fundamental structures: an array of thin wires (to create an electric dipole response) and an array of SRRs (to generate a magnetic dipole response).

Over the last two decades, the radiation pattern of antennas has been controlled by placing metamaterials in the near-field region of conventional antennas and provide wide beam-steering and beam-forming capabilities [7–8]. When graded-index metamaterials are integrated into antennas, they enable higher deflection angles while minimally affecting other antenna properties. Reconfigurable metamaterial-loaded antennas offer superior multidirectional beam deflection [9]. Extraordinary properties of metamaterials, such as negative permittivity, negative permeability, and negative refractive index, arise from their geometric configurations rather than their chemical composition. This makes them distinct from natural materials and offers more flexibility to tune performance compared to conventional materials [10]. This capability allows engineers to design advanced RF devices with greater freedom and efficiency [11].

Lu et al. [12]designed a PRA where complementary SRRs (CSRRs) were integrated into the ground plane of an antenna to enable directional control of the radiation pattern across nine angles. The patch antenna was coupled with four CSRR cells, which were modified using slots and PIN diodes to achieve variable configurations. Similar work was reported in Ref. [13].

Rachedi et al. [14] proposed a compact antenna design capable of dynamic beam steering by leveraging the unique properties of SRRs, when a simple SRR placed in proximity to a double SRR, excited by an RF voltage source, resulted in an omnidirectional radiation pattern. As depicted in Figure 3.6, a parasitic SRR placed in the immediate vicinity of the main SRR behaved as a magneto-electric resonator, deflecting the main beam.

Metamaterials have proven useful in high-frequency antenna designs [15–16]. For example, pattern reconfiguration was achieved using a microstrip-fed patch antenna with an SRR metamaterial layer, as shown in Figure 3.7(a)–(b) [16]. When a metamaterial structure is placed in the near-field region of an electromagnetic source, it acts as a phase transformation device. By replacing complex phased-array structures with metamaterial layers, wide-range 2D beam steering can be realized. The radiation pattern of the antenna in the azimuth and elevation planes, both with and without the

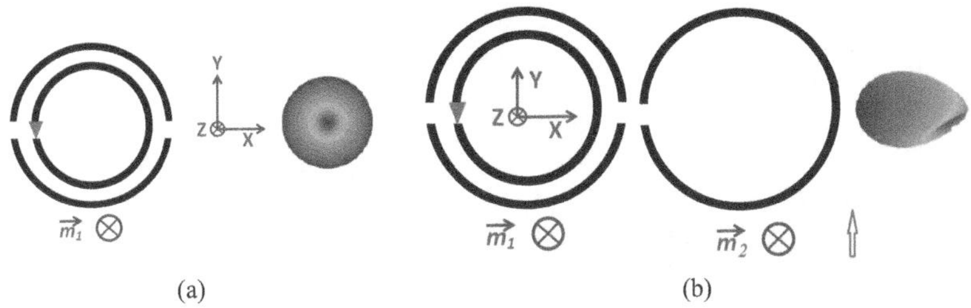

Figure 3.6 (a) Geometry and radiation pattern of the double Split Ring Resonator (SRR). The red triangle represents the voltage source (b) Coupled resonator structure its radiation pattern.

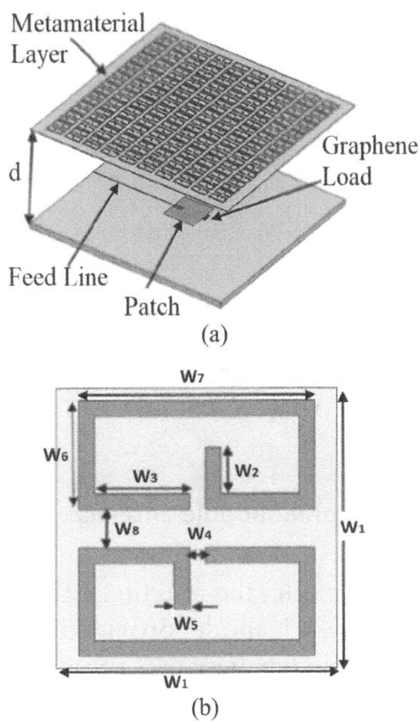

Figure 3.7 (a) The geometry of the prototype antenna (b) dimensions of metamaterial unit cell. The radiation pattern of the prototype antenna for phi=0 and 90 at 1 THz.

metamaterial layer. Notably, the presence of the metamaterial layer enhances antenna gain and improves the front-to-back ratio despite the higher backlobe levels.

Yan and Vandenbosch [17], developed a metamaterial-loaded pattern-reconfigurable antenna operating at 2.4 GHz, designed for on-body applications, primarily in ISM bands as shown in Figure 3.8. Pattern reconfiguration was achieved by modifying the dispersion curve of the transmission line with a metamaterial layer, enabling the antenna to resonate in 0th and +1 mode.

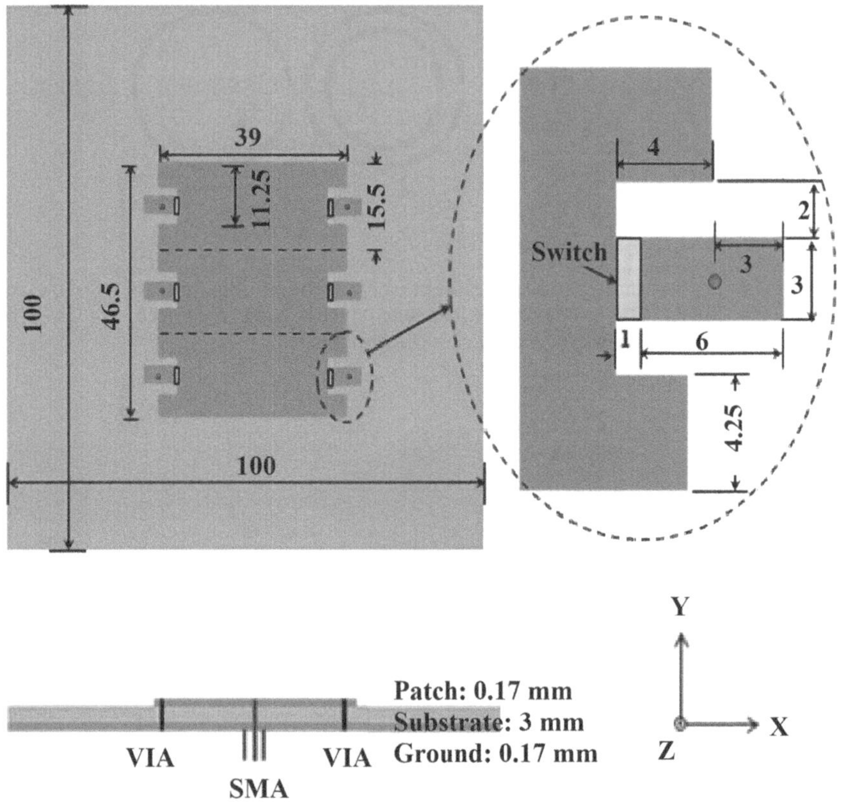

Figure 3.8 Antenna topology of monopole antenna.

When operating in the 0th mode, the antenna exhibited an omnidirectional radiation pattern while the +1 mode provided a unidirectional pattern. A PIN diode was used to switch between these modes. In reference [18], Esmil et al. proposed a dipole antenna loaded with a combination of contiguous square resonator and three switches to provide two different metamaterial of different refractive indices which deflected the antenna radiation beam. This antenna provided a main beam deflection of +30° and −27° in the E-plane at 28 GHz. Such designs highlight the versatility of metamaterial-based reconfigurable antennas in wireless communication systems.

In the case of epsilon-near-zero (ENZ) materials, their permittivity is near zero or negative, resulting in a nearly infinite wavelength within the material. These properties enable electromagnetic waves to propagate with minimal phase variation, facilitating effective wave confinement and redirection. This makes it easier to dynamically steer the antenna's radiation pattern or focus energy in a desired direction. Wang and Dong [19] presented a compact antenna design leveraging ENZ materials to achieve pattern reconfigurability without compromising performance. This design enhances energy efficiency by reducing losses, making it highly suitable for IoT communication systems. On the other hand, MNZI metamaterial is an engineered material with magnetic permeability that is negative and close to zero at specific frequencies. When the permeability of such material approaches zero, the magnetic response becomes extremely weak, allowing electromagnetic waves to propagate through it with minimal interaction with the magnetic field. This results in electromagnetic waves experiencing minimal phase variation as they travel through the material. MNZI metamaterials exhibit unique electromagnetic properties, enabling exceptional control over wave propagation, impedance matching, and radiation characteristics. A switchable MNZI metamaterial that facilitates dynamic beam-scanning lens antenna reconfiguration using a single feed was developed by [20]. This material, with its low refractive index and magnetic tunability, provides efficient and precise beam steering, addressing challenges in compact and adaptive antenna systems.

SRRs offer a compact design with a strong magnetic response, enhancing beam-steering capabilities. They are particularly useful for narrowband applications due to their sharp resonance, but they have limited dynamic reconfigurability, as reconfiguration often requires physical modifications. Additionally, SRRs can experience significant losses at higher frequencies.

Caloz and Itoh [21] explored the use of metamaterials and metasurfaces in designing beam-steering and beam-shaping PRAs by arranging ESRPs in specific configurations. These configurations alter the antenna's directionality, gain, and beam shape dynamically by incorporating PIN diodes into the metamaterial, allowing sections of the structure to activate or deactivate as needed. Similarly, Alù and Engheta[22] proposed a design where PIN diodes or varactors are integrated within metamaterials to enable dynamic control over the antenna's radiation pattern by activating or deactivating SRRs.Huang and Sarabandi[23] developed a varactor-tuned (voltage-controlled capacitor) metamaterial-based compact PRA. This innovation makes the design suitable for mobile communications, IoT devices, and other systems where size, flexibility, and efficiency are critical. The integration of varactors helps in size reduction, enhances pattern reconfigurability, and contributes significantly to the advancement of adaptive and compact antenna technology.

FSSs are planar periodic arrays of conductive elements on a dielectric substrate that allow or block specific frequencies, acting as spatial filters with wide-ranging applications in wireless communication systems. Different unit cell element shapes and geometries, such as rings, fractal slots, or square patch can be used to achieve various responses.

In the past few years, researchers have proposed various PRA designs using FSS structures and geometrical configurations, combined with different electronic switching devices to control the surface frequency response of the antenna by keeping certain parts of the antenna switched between transparent and reflective states. FSS structures are typically thin and lightweight, ensuring that the overall antenna remains compact. Additionally, electronic switches can rapidly switch between states, enabling real-time pattern reconfiguration. The integration of electronic switches within the FSS provides flexibility in controlling radiation patterns, making these setups adaptable to diverse environments and application requirements. In Ref. [24] (2021), FSSs are highlighted as being extensively used to improve antenna performance. They can also be combined with tunable components to enable pattern reconfigurability over multiple bands. FSSs are effective for both transmissive and reflective designs, making them versatile.

For instance, a monopole radiating source surrounded by four columns of an FSS screen and artificial medium, where each unit cell consists of two metallic crosses connected by a PIN diode, facilitates beam switching in the azimuth plane at 2.5 GHz by activate or deactivate the PIN diodes [25]. Similarly, in Ref. [26] (2021), the authors focused on the unique properties of hexagonal geometries to improve the performance and bandwidth of FSSs compared to conventional rectangular or square grids. These hexagonal interwoven FSS structures can be applied in antenna designs to control radiation patterns, enhance gain, and suppress unwanted side lobes.

In Ref. [27], FSS scatterers consist of capacitive and inductive elements, which generate a bandpass for the incident electromagnetic waves. The integration of varactor diodes tunes the capacitance of the LC circuit in the FSS, enabling control over the bandpass frequency. An FSS unit cell composed of metallic discontinuous strips with PIN diodes at their discontinuities can be embedded cylindrically around a dipole antenna with an omnidirectional radiation pattern [28]. The FSS configuration comprises two semi-cylinders connected by PIN diodes. By changing the states of the PIN diodes, the radiated beam of the antenna can be steered throughout the entire azimuth plane. 1n 2012, Ganesh et al. [29], demonstrates an antenna with two strategically placed FSS layers. One acts as a reflector and the other as a director, enabling beam steering in three directions (265°, 13°, and 337°) on the azimuth plane.

In summary, by selectively activating different elements of the FSS, the surface can focus or redirect radiated energy in different directions, providing beam-steering capabilities. FSS structures can allow an antenna to switch between omnidirectional and directional radiation patterns by turning certain sections on or off. They can also be designed to allow specific frequency bands to pass through or reflect at various angles. Adjusting the states of PIN diodes enables antennas to operate at multiple frequencies or modes.

3.2.3 HISs or AMCs

HISs, also known as AMCs, are specialized materials designed to control electromagnetic waves by achieving specific impedance values that typically approach infinity. This property prevents wave propagation in certain directions while allowing it in others. HIS effectively suppresses the propagation of surface waves, which are unwanted waves traveling along the surface of conductors or dielectric materials, thus enhancing overall antenna performance. These surfaces are also beneficial in shielding applications to prevent electromagnetic interference, particularly in high-frequency circuits.

HIS structures are engineered to reflect electromagnetic waves with a specific phase shift. They are often composed of periodic square or circular metallic patches or mushroom-like structures placed on a dielectric material to form the HIS. The periodic array of patches helps reflect incident waves, control propagation in desired directions, improve antenna efficiency, and suppress side lobes. HIS structures are particularly useful for controlling reflection phase, making them valuable for pattern shaping and beam steering. They are also highly tunable when combined with components like varactors or PIN diodes, allowing for dynamic reconfigurability.

However, HIS structures are complex to fabricate, especially when fine-tuning components are included. Additionally, their inherent frequency-selective nature and narrowband characteristics limit their use in wideband or multiband antenna applications. HIS structures exhibit negative magnetic permeability, which is why they are also referred to as AMCs. In Ref. [30], a perfect electric conductor (PEC) reflects incident electromagnetic waves with a reflection phase of 180°, whereas a PMC produces a reflection with zero phase shifts. However, PMCs do not exist in nature. AMCs approximate PMCs but have limited bandwidth and cross zero reflection phase at only one frequency point. The effective bandwidth of an AMC typically lies within a range of ±90° degrees around the center frequency. Unlike FSS, AMCs have a ground plane positioned behind periodic structure. Deo et al. [31], proposed an HIS integrated antenna, where beam steering is achieved by different combinations of active and inactive four open-circuit switches over an HIS layer. The designed antenna operates at 3.3 GHz and has an average directivity 6.9 dBi. The main lobe of the antenna can switch between 16 different positions, covering nearly 360 beam steering.

In Ref. [32], reconfigurable AMCs, designed using capacitively loaded loops (CLLs), act as reflective surfaces to enhance antenna gain and control radiation patterns. Reconfigurable AMC elements allow the antenna to dynamically adjust its operating frequency and directionality, which is valuable for multi-band and adaptive systems. These adjustments are achieved using electronic components such as RF MEMS or PIN diodes, enabling real-time changes without altering the physical structure.

In Ref. [33], a PRA is composed of an E-shaped dipole and an active AMC layer with a 5x5 periodic array. Each AMC unit cell includes a compound fractal ring loaded with two PIN diodes. Pattern reconfiguration is achieved by toggling the state of the PIN diodes in different rows. The proposed PRA switches between ±39° in the elevation plane. Lago et al. [34], proposed an antenna integrating AMC structures (Figure 3.9), parasitic elements, and RF MEMS switches for reconfigurable beamforming. The folded dipole antenna provides increased bandwidth and improved impedance matching. The AMC layer reduces ground plane interaction, improving radiation patterns and directivity. RF MEMS switches, preferred for their low power consumption and high performance in high-frequency applications, allow for dynamic beam direction adjustments up to ±58 degree.

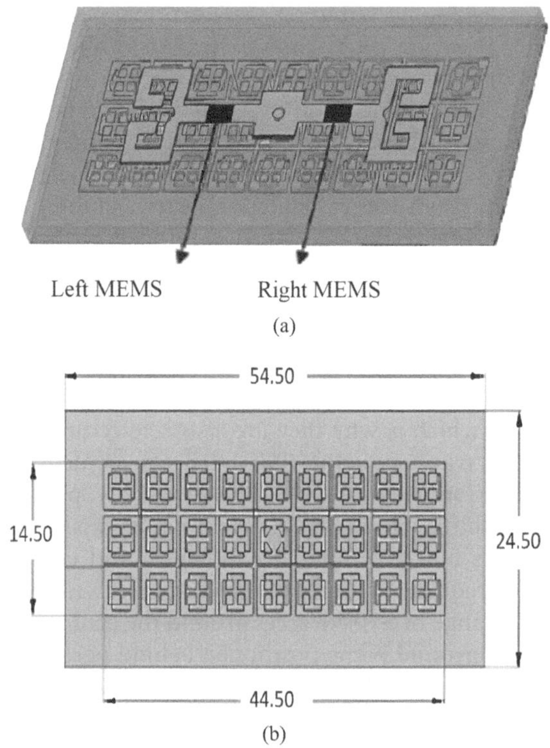

Figure 3.9 Proposed antenna (a) 3-D view, (b) AMC plane [34].

3.2.4 Electromagnetic band gap (EBG) structures

EBG structures are engineered materials designed to control electromagnetic behavior in ways not typically found in natural materials, often considered a subset of metamaterials. EBG structures are periodic arrangements that prevent electromagnetic wave propagation within certain frequency bands, suppress surface waves, and reduce coupling between antennas, thereby improving performance in high-frequency applications. EBG structures operate on the principle that their unit cells are smaller than the wavelength of the electromagnetic waves they are designed to affect. This similarity places EBGs under the metamaterial umbrella. In Refs. [35–37], researchers explored EBG structures for radiation pattern control. In 2019, Abdulhameed et al [37],developed two novel antennas using EBG structures to achieved pattern reconfiguration by inserting or removing vias, with EBG positioned under the antenna on a double substrate. Habib et al. [38], present the design of a beam-switching antenna operating at 1.8 GHz and a gain of 10 dBi was designed. The antenna switched between six beams with a 60° beam width, covering 360° in the azimuth plane.

3.2.5 Metasurfaces

Metasurfaces are 2D artificial materials designed to manipulate electromagnetic waves at a sub-wavelength scale. They consist of thin layers of nanoscale structures arranged in specific patterns to achieve unique electromagnetic properties, allowing precise control over wavefronts, polarization, phase, and amplitude. The meta-atoms on a metasurface, interact with incoming electromagnetic waves to induce phase shifts, reflections, or transmissions. This interaction is determined by the design of the meta-atoms, enabling the metasurface to focus, deflect, or shape waves as needed. Metasurfaces are ultra-thin, compact, and lightweight, making them suitable for designing PRAs for various wireless communication services. A single and double layer metasurface loaded antenna is shown in Figure 3.10 [39].

In 2019 Chaimool et al. [40], explores the development of a metasurface antenna utilizing PIN diodes to achieve dynamic beam steering capabilities. Periodic or quasi-periodic metallic elements form the metasurface. The metasurface changes its behavior as reflector (behaves as ENG) or director (behaves MNZ) when the PIN diode toggles its status. The proposed antenna

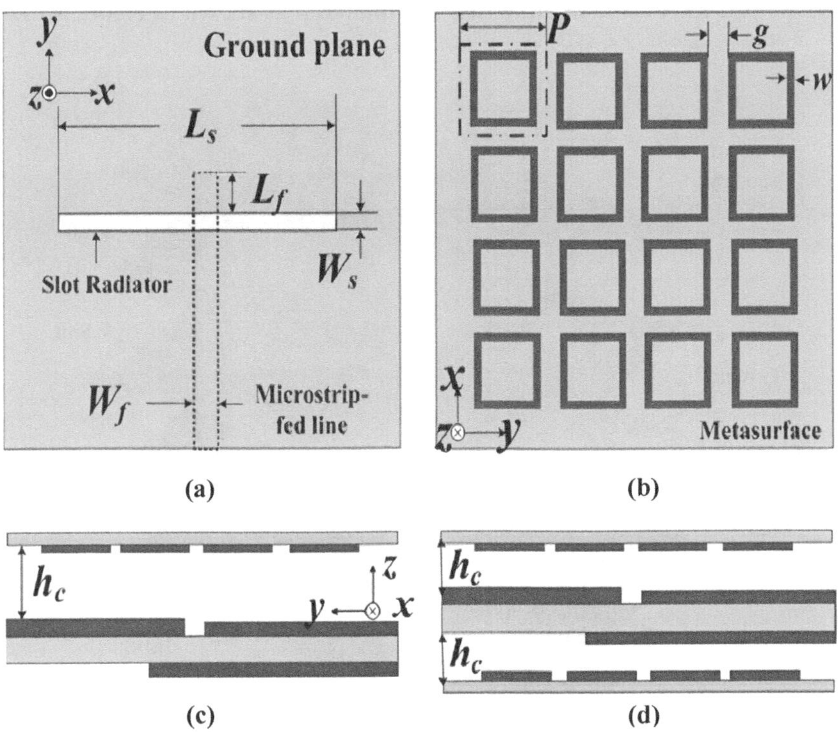

Figure 3.10 (a) Microstrip fed slotted antenna, (b) a square ring metasurface, (c) antenna loaded with metasurface layer, (d) antenna with two layers metasurface [39].

enables beam steering at ±30° and ±60° with high accuracy, good impedance matching, and stable gain across the operational frequency range, making it suitable for 5G networks, radar systems, and mobile communications.

In Ref. [41], a wideband monopole antenna is combined with a programmable metasurface acting as a reflector. The metasurface consists of a 4x4 array of square ring-shaped unit cells, each controlled by diodes. This setup enables precise manipulation of the radiation pattern and beam direction. The antenna can switch between omnidirectional and directional radiation patterns, as well as achieve beam steering in both elevation and azimuth planes.

3.2.6 Liquid crystal-based metamaterials

Research on liquid crystal-based metamaterials, such as [42–44], demonstrates their potential for antenna reconfiguration. By applying external electric fields, the orientation of liquid crystal molecules in the metasurface can be dynamically adjusted, altering its electromagnetic properties. This allows for continuous tuning of the radiation pattern and beam direction. Liquid crystal technology provides low-power operation and fast switching, offering flexibility in controlling antenna performance. Zhao et al. [43], developed PRA using liquid based metamaterial as shown in Figure 3.11.

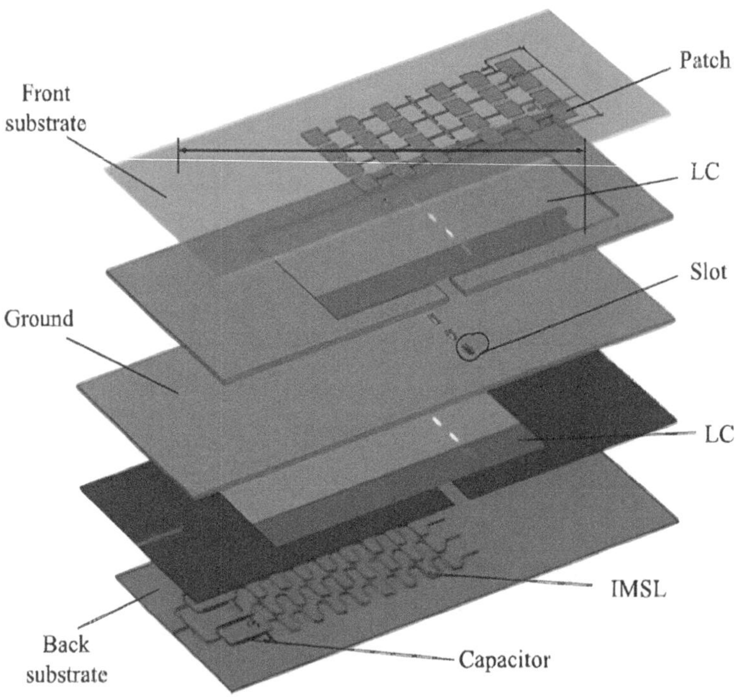

Figure 3.11 Schematic model of the proposed reconfigurable antenna array [43].

3.2.7 Graphene-based metamaterials

Graphene-based metamaterials leverage graphene's conductive and tunable properties, which can be dynamically adjusted through an external bias. These materials offer ultra-fast tunability and broad bandwidth capabilities, particularly in the terahertz (THz) frequency spectrum, where traditional materials and switches are less effective. Their thin and flexible nature makes them attractive for modern wireless communication devices. Such metamaterials are expensive and require precise control of external bias to achieve desired tuning and are still in early stages of development, with fabrication challenges. The conductivity of graphene changes with chemical potentials regulated by electrostatic bias voltages, enabling dynamic control of antenna's resonance frequency and radiation pattern.

In Ref. [45], Xu et al. proposed a graphene-based MIMO antenna system for THz communication. An array of graphene layers was used to control radiation direction and beam width by adjusting the chemical potential through bias voltage, enabling beam deflection up to 80°.Wang et al., in Ref. [46], introduced a 2D leaky-wave antenna employing a graphene-based HIS. Adjusting graphene's conductivity allowed reconfiguration of the reflection phase, enabling frequency tuning and beam steering up to 68°. In Ref. [47], Li et al. designed a graphene-based leaky-wave antenna operating in the THz band. By varying the applied bias voltage on graphene, the surface plasmon polaritons were controlled, achieving a beam scanning range of up to 75°. Luo et al., in Ref. [48], explored graphene's use in creating a negative refractive index metamaterial for THz frequencies. The structured graphene layers enabled dynamic beam tilting up to 27°.The study [49] highlighted graphene's ability to tune antenna radiation patterns. The design utilized a square spiral structure with reconfigurability achieved through chemical potential adjustments, enabling operation across a broad frequency range (0.1–2 THz). Despite the significant advancements in graphene-based antennas for THz band applications, many designs exhibit limited beam steering ranges and low radiation efficiencies. These limitations fall short of meeting the performance demands of THz communication systems, which require high efficiency and broader steering capabilities.

3.2.8 Summary table

Each metamaterial structure offers distinct advantages depending on the requirements, such as reconfigurability speed, beam scanning range, frequency range, and application environment. By choosing the right structure, designers can optimize pattern reconfigurability, efficiency, and flexibility for specific antenna applications. Table 3.1 summarizes the beam defection antenna using different types of MMs.

Table 3.1 Summary of the beam-scanning antenna using different types of MMs

ref	*Antenna type*	*Frequency band*	*gain*	*Beam deflection angle (degree)*	*Reconfi-gurable method*	*Metamaterial structure*
[24]	Dipole antenna	2.45 GHz	—	Switching with 20° steps	Varactor diode	Cylindrical FSS
[25]	Monopole antenna	5.2 GHz	—		PIN diode	4 column of FSS
[29]	Hexagonal patch	6.5 GHz	Enhanced by 2–4 dBi	265°, 13°, and 337°	PIN diode	FSS
[31]	Rectangular spiral antenna	3.3 GHz	6.9 dBi	16 switching options	PIN diode	HIS
[32]	Monopole antenna	Sub 6GHz band	4.9 dBi	12 switching potions separated by 30°	PIN diode/ RF MEMS	AMC
[33]	E-shaped dipole antenna	2.68 GHz	7.84 dBi	±39°	PIN diode	Compound fractal ring AMC
[34]	Folded dipole antenna	9.41 GHz	8.08 dBi	Upto ±58°	MEMS switches	AMC
[37]	Rectangular antenna	6 GHz	9.86 dB	E-plane ±20°H plane ±18°	—	EBG
			10.16 dB	E-plane ±17°H-plane + ±15°	—	EBG
[38]	Monopole antenna	1.8 GHz	10 dBi	6 beams with a 60° beamwidth	—	EBG Metamaterial using FSS
[40]	Monopole antenna	2.45 GHz		±30° and ±60°	PIN diodes	Metasurface
[41]	Monopole antenna	5.8 GHz	8.89 dBi	+13° (xz-elevation) ±10° (yz-elevation)	PIN diodes	Metasurface (4x4 square ring)

3.3 CHALLENGES AND LIMITATIONS

Design and development of an efficient and cost-effective PRA is always a challenge for antenna designers. High gain, better efficiency, a stable radiation pattern and impedance matching for a complete frequency band are a few of the important factors to consider at the time of antenna design. Researchers found that metamaterial based PRAs offer several advantages, but they have some inherent limitations like saturation of resonance frequency and limited bandwidth. Most of the designs for PRAs available in literature focused to achieve desired characteristics by compromising with other performance characteristics. These limitations can be overcome by incorporating tunable elements into metamaterial based devices. On the other hand, integration of switches and its biasing circuitry makes the system more complex and less efficient. It affects the deflection angle and hence degrades the radiation performance of a PRA. The gain of the antenna can be improved by increasing array size of the metamaterial (more meta cells) but this also increases the beam width of the antenna radiation pattern which effectively reduces the directivity of the antenna.

For better performance of metamaterial based antenna, we should have a good understanding of factors like, metamaterial structure, position of unit cells, and feeding mechanism etc. which till date is very limited. With the help of theoretical formulation of this type of antenna, it will help a lot in understanding the main parameters of antenna. Besides, it will also provide design guidelines to antenna designers. It is clear from the available literature that this field is not yet mature and much remains to be learned about it. In the past years, researchers have done a lot of work in this field and there are many good examples in the literature which help a lot in developing metamaterial based PRA. Despite this, there is very little scientific work in this field and most of the designs have been developed on the basis of electromagnetic simulators.

3.4 CONCLUSION

In this paper, the gist of various research articles published in the last two decades related to the design and development of metamaterials inspired PRMAs has been compiled for better understanding of the subject for antenna designers. Many such methods have been mentioned by which the radiation pattern of a microstrip antenna can be controlled in the desired manner by using different types of metamaterials. Implementation of metamaterials or metasurfaces with traditional methods to design PRA can be fabricated in an even better and easier way. Inserting metamaterials in the vicinity of the radiating element of a conventional microstrip antenna manipulates the antenna's surface current, which results in the pattern reconfiguration of the original antenna. Metamaterials also help to develop more compact, larger bandwidth and continued steering antenna.

REFERENCES

[1] J. Kowalewski, J. Atuegwu, J. Mayer, T. Mahler, and T. Zwick, "A Low-Profile Pattern Reconfigurable Antenna System for Automotive MIMO Applications," *Progress In Electromagnetics Research*, vol. 161, pp. 41–55, 2018.

[2] L. Reichardt, T. Mahler, Y. L. Sit and T. Zwick, "Using a synthesis methodology for the design of automotive antenna systems," *7th European Conference on Antennas and Propagation (EuCAP)*, Gothenburg, Sweden, pp. 1600–1604, 2013.

[3] J. B. Mandelstam, "Group velocity in crystal lattice," *Journal of Experimental and Theoretical Physics*, vol. 15 (9), pp. 475–478, 1945.

[4] V. G. Veselago, "The electrodynamics of substances with simultaneously negative values of epsilon and μ," *Soviet Physics Uspekhi*, vol. 10, pp. 509–514, 1968.

[5] D. R. Smith, W. Padilla, D. Vier, S. Nemat-Nasser, S. Schultz, "Composite medium with simultaneously negative permeability and permittivity," *Physical Review Letters*, vol. 84 (18), pp. 4184–4187, 2000.

[6] B. Virdee, "Grand challenges in metamaterial antennas," *Frontiers in Antennas and Propagation*, 2022. https://doi.org/10.3389/fanpr.2022.1032205

[7] K. Payne, J. H. Choi, "Development of dual-band/dual-polarized metamaterial antenna in an electrically-small form factor for efficient wireless data and power transfer," *IEEE Conference on Antenna Measurements and Applications, CAMA 2016*, 7815788, 2016.

[8] E. Torabi, D. Erricolo, P. -Y. Chen, W. Fuscaldo and R. Beccherelli, "Wide beam steering by leaky wave antenna with liquid crystal-based phasing structure and metamaterial-based radiating apertures in Ku band," *IEEE International Symposium on Antennas and Propagation and USNC-URSI Radio Science Meeting (AP-S/URSI)*, Denver, CO, USA, pp. 1586–1587, 2022.

[9] B. A. F. Esmail, S. Koziel, L. Golunski, H. B. A. Majid and R. K. Barik, "Overview of metamaterials-integrated antennas for beam manipulation applications: The two decades of progress," *IEEE Access*, vol. 10, pp. 67096–67116, 2022.

[10] R. Mark, N. Rajak, K. Mandal, S. Das, "Metamaterial based superstrate towards the isolation and gain enhancement of MIMO antenna for WLAN application," *AEU-International Journal of Electronics and Communications* Vol. 100, pp. 144–152, 2019.

[11] H. A. Majid, M. K. A. Rahim, T. Masri, "Left handed metamaterial design for microstrip antenna application," *Proceedings of the IEEE International RF and Microwave Conference*, Kuala Lumpur, Malaysia, pp. 2–4, 2008.

[12] Z. L. Lu, X. X. Yang, and G. N. Tan, "A multidirectional pattern-reconfigurable patch antenna with CSRR on the ground," *IEEE Antennas and Wireless Propagation Letters*, vol. 16, pp. 1–11, 2016.

[13] U. R. Khan, A. Basit, J. Shekh, G. M. Bhat, S. Ahmad, "Design of multiband pattern reconfigurable antenna loaded with circular split ring resonators," *Recent Innovations in Computing*, Springer, pp. 315–326, 2022.

[14] K. Rachedi, J. Rosny, Y. Kokar, D. T. P. Huy, and A. Ourir, "Beam steering reconfigurable compact antenna based on hybridization between split ring resonators," *Progress in Electromagnetics Research M*, vol. 91, pp. 189–196, 2020.

[15] A. K. Singh, M. P. Abegaonkar, S. K. Koul, "High-gain and high-aperture-efficiency cavity resonator antenna using metamaterial superstrate," *IEEE Antennas and Wireless Propagation Letters*, vol. 16, pp. 2388–2391, 2017.

[16] M. M. Seyedsharbaty and R. A. Sadeghzadeh, "Antenna gain enhancement by using metamaterial radome at THz band with reconfigurable characteristics based on graphene load," *Optical and Quantum Electronics*, vol. 49, pp. 1–13, 2017.

[17] S. Yan and G. A. E. Vandenbosch, "Radiation pattern-reconfigurable wearable antenna based on metamaterial structure," *IEEE Antennas and Wireless Propagation Letters*, vol. 15, pp. 1715–1718, 2016.

[18] E. Smail et al., "Reconfigurable radiation pattern of planar antenna using metamaterial for 5G applications," *Materials*, vol. *13*(3), pp. 582, 2020.

[19] Z. Wang and Y. Dong, "Novel pattern reconfigurable epsilon-near-zero (ENZ) antenna for intelligent IoT communication applications," *IEEE Internet of Things Journal*, vol. 11, no. 12, pp. 22364–22375, 2024.

[20] J. P. Turpin and D. H. Werner, "Switchable near-zero-index magnetic metamaterial for dynamic beam-scanning lens," *Antennas and Propagation Society International Symposium (APSURSI)*, 2012. https://doi.org/10.1109/APS.2012.6348076

[21] C. Caloz and T. Itoh, "Array factor approach of leaky-wave antennas and application to 1-D/2-D composite right/left-handed (CRLH) structures," *IEEE Microwave and Wireless Components Letters*, vol. 14(6), pp. 274–276, 2004.

[22] A. Alù and N. Engheta, "Microwave Metamaterials: Selected Features and Sample Applications," *Tutorials in Metamaterials*, pp. 73–98, 2016.

[23] Sarabandi, "A varactor-tuned dual-band slot antenna," *IEEE Transactions on Antennas and Propagation*, vol. 54, pp. 401–408, 2006.

[24] L.-M. Zhang, X. Ding and W. Shao, "A phase compensation beam switching antenna based on frequency selective surface," *IEEE Antennas and Wireless Propagation Letters*, vol. 20, no. 9, pp. 1741–1744, 2021.

[25] J. Li, Q. Zeng and T. A. Denidni, "Pattern reconfigurable antenna loaded with frequency selective surface and artificial dielectric medium," *2018 IEEE International Symposium on Antennas and Propagation & USNC/URSI National Radio Science Meeting*, Boston, MA, USA, pp. 975–976, 2018, 2018.

[26] J. A. Vasquez-Peralvo, J. M. Fernandez-Gonzalez, and J. M. Rigelsford, "Frequency selective surfaces using hexagonal interwoven structures," *Proceedings of ISAP2020, Osaka, Japan*, pp. 265–266, 2021.

[27] M. Sazegar et al., "Beam steering transmitarray using tunable frequency selective surface with integrated ferroelectric varactors," *IEEE Transactions on Antennas and Propagation*, vol. 60, no. 12, pp. 5690–5699, 2012.

[28] A. Edalati and T. A. Denidni, "High-gain reconfigurable sectoral antenna using an active cylindrical FSS structure," *IEEE Transactions on Antennas and Propagation*, vol. 59, no. 7, pp. 2464–2472, 2011.

[29] M. Ganesh, N. S. Raghava, T. Sabapathy and Y. Sharma, "A gain enhanced multiband frequency and pattern reconfigurable antenna for Wi-Fi 6E and 5G new radio wireless standards," *International Journal of Communication Systems*, 2024, https://doi.org/10.1002/dac.6011

[30] H. Zhou and F. Xu, "Artificial magnetic conductor and its application," *2013 Proceedings of the International Symposium on Antennas & Propagation*, Nanjing, China, pp. 1110–1113, 2013.

[31] P. Deo, A. Mehta, D. Mirshekar-Syahkal and H. Nakano, "An HIS-based spiral antenna for pattern reconfigurable applications," *IEEE Antennas and Wireless Propagation Letters*, vol. 8, pp. 196–199, 2009.

[32] Singh, Vikrant et al., "A metasurface-based electronically steerable compact antenna system with reconfigurable artificial magnetic conductor reflector elements," *iScience*, vol. 25, 2022, pp. 1–17.
[33] Y.-X. Lin, W.-P. Cao, L.-L. Sheng and S.-M. Li, "Design of a high gain pattern-reconfigurable antenna using active AMC surface," *2021 International Conference on Microwave and Millimeter Wave Technology (ICMMT)*, Nanjing, China, pp. 1–3, 2021.
[34] L. Herwansyah et al. "AMC-integrated reconfigurable beamforming folded dipole antenna with parasitic and RF MEMS." *Progress in Electromagnetics Research C*, vol. 69, pp. 159–167, 2016.
[35] I. Muhammad Faizal et al. "Pattern reconfigurable antenna using electromagnetic band gap structure." *Applied Physics A*, vol. 123, pp. 1–5, 2016.
[36] H. Omar, "Dual band pattern reconfigurable antenna using electromagnetic band gap structures," *AEU - International Journal of Electronics and Communications*, vol. 130, pp. 1–10, 2021.
[37] M. K. Abdulhameed, M. B. M. Isa, Z. Zakaria, I. M. Ibrahim, M. K. Mohsen, M. L. Attiah, A. M. Dinar, "Radiation control of microstrip patch antenna by using electromagnetic band gap," *AEU - International Journal of Electronics and Communications*, vol. 110, pp. 1–11, 2019.
[38] M. A. Habib, M. N. Jazi, A. Djaiz, M. Nedil and T. A. Denidni, "Switched-beam antenna based on EBG periodic structures," *IEEE MTT-S International Microwave Symposium Digest*, Boston, MA, USA, pp. 813–816, 2009.
[39] S. Chaimool, C. Rakluea, and P. Akkaraekthalin, "Mu-near-zero metasurface for microstrip-fed slot antennas" *Applied Physics A*, vol. 112, pp. 669–675, 2013.
[40] S. Chaimool, T. Hongnara, C. Rakluea, P. Akkaraekthalin, Y. Zhao, "Design of a PIN diode based reconfigurable metasurface antenna for beam switching applications," *International Journal of Antennas and Propagation*, vol. 2019, pp. 1–7.
[41] W. A. Awan, N. Hussain, S. G. Park, N. Kim, "Intelligent metasurface based antenna with pattern and beam reconfigurability for internet of things applications," *Alexandria Engineering Journal*, vol. 92, pp. 50–62, 2024.
[42] R. Guirado, G. P. Palomino, M. Ferreras, E. Carrasco, and M. Cano-Garcıa, "Dynamic modelling of liquid crystal based metasurfaces and its application to reducing reconfigurability times," *IEEE Transactions on Antennas and Propagation*, vol. 70, pp. 11847–11857, 2022.
[43] Z. Yizhe et al. "A frequency and pattern reconfigurable antenn based on Antenna Array Based on Liquid Crystal Technology." *IEEE Photonics Journal*, vol. 9, pp. 1–7, 2017.
[44] L. Liu, L. Deng, S. Li, C. Zhang, L. Wang and M. Qu, "Beam reconfigurable antenna based on liquid crystal metasurface," *IEEE 4th International Conference on Computer and Communications (ICCC)*, Chengdu, China, pp. 1037–1041, 2018.
[45] Z. Xu, X. Dong and J. Bornemann, "Design of a reconfigurable MIMO system for THz communications based on graphene antennas," *IEEE Trans Terahertz SciTechnol*, vol. 4 (5), pp. 609–617, 2014.
[46] X. Wang, W. Zhao, J. Hu, and W. Y. Yin, "Reconfigurable Terahertz leaky-wave antenna using graphene-based high-impedance surface," *IEEE Transactions on Nanotechnology*, vol. 14, pp. 62–69, 2015.

[47] J. Li, M. He, and C. Wu, et al., "Radiation pattern reconfigurable graphene leaky-wave antenna at terahertz band based on dielectric grating structure," *IEEE Antennas and Wireless Propagation Letters*, vol. 16, pp. 1771–1775, 2017.

[48] L. Yanbin et al. "A graphene-based tunable negative refractive index metamaterial and its application in dynamic beam-tilting terahertz antenna," *Microwave and Optical Technology Letters*, vol. 61, pp. 2766–2772, 2019.

[49] X. Zhang, C. Ruan and J. Dai, "Reconfigurable Antenna Based on Graphene at Terahertz Frequency," *2018 Progress in Electromagnetics Research Symposium (PIERS-Toyama)*, Toyama, Japan, pp. 2311–2313, 2018.

Chapter 4

Computer intelligence in health care field

Transforming the practice of medical applications

Sukhdeep Kaur, Kawerinder Singh Sidhu, Konda Hari Krishna, Kapil Joshi, Ajay Singh, and Nadia Mahmood Hussien

4.1 INTRODUCTION

Over the course of three millennia, the study of (human) intelligence has developed a rich history. The development of computers in the 20th century made it possible to create and research systems that display traits or actions typically associated with intelligence, but that are not natural in the sense that they were created by humans. The field commonly referred to as artificial intelligence is an engineering correction or emerging discipline.

The idea, conception, use, and progression of computing patterns determined by linguistic and biological factors is known as computational intelligence. Historically, evolutionary computation, neural networks, and fuzzy systems have been the three primary foundations of computational intelligence. As a result, computational intelligence is a developing area that currently includes computing architypes such as, ambient intelligence, artificial life, artistic learning, non-natural endocrine networks, societal reasoning, and artificial hormone networks in accumulation to its three primary components. The development of successful intelligent systems, such as games and cognitive developmental systems, heavily relies on computational intelligence. Research on deep learning, namely deep convolutional neural networks, has exploded in the last few years. Deep learning is now the main technique used in artificial intelligence. In actuality, computational intelligence is the foundation of some of the most effective artificial intelligence systems. Figure 4.1 shows the foundation of computer intelligence.

4.1.1 Fuzzy systems

Created on a generality of out-of-date reason, fuzzy systems (FS) solve uncertain problems and model linguistic imprecision by drawing inspiration from human language. This allows us to perform approximate reasoning.

 DOI: 10.1201/9781003568995-4

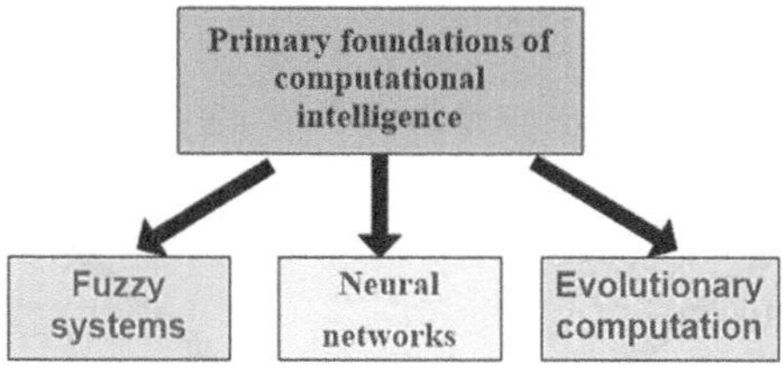

Figure 4.1 Foundations of CI.

4.1.2 Neural networks

Artificial neural networks are immensely similar distributed networks that can generalize and learn from instances, drawing inspiration from the social brain.

4.1.3 Evolutionary computation

Evolutionary computation, which entices stimulation from genetic evolution, generates, assesses, and adjusts a population of potential solutions to optimize problems.

The revolution of persistent care facilities is largely attributed to the computerization of digital technologies, mostly in the healthcare engineering [1]. Due to the sophistication of healthcare systems, which makes it difficult to achieve effective operations, digital transformation is now essential for innovation and progress. Artificial intelligence, one of the most important and groundbreaking advances in modern computer science, has the potential to transform many different sectors and methods of operation. Because of the technology's potential influence, particularly in terms of automation, academics and policymakers are nevertheless interested in artificial intelligence in the healthcare sector [2]. Therefore, it is essential to ascertain how artificial intelligence will affect the automation of digital health system technology.

Digital health schemes aim to update and improve the distribution of healthcare facilities, making it possible to address ingrained inefficiencies that obstruct high-quality care on a number of levels [3]. AI has revolutionized the healthcare sector due to its capacity to investigate enormous quantities of data, identify patterns, and generate insights more quickly than humans. Machine learning algorithms, a subset of artificial intelligence, are crucial for evaluating sizable, complex datasets and giving medical practitioners pertinent information to aid in their decision-making. From early diagnosis to individualized treatment, AI's capabilities are transforming the conventional medical scene [4]. Healthcare benefactors typically have admittance to a sophisticated and perceptible framework through the use of digital technologies, which empowers consumers by enabling choice in an environment where health service-related information is more actionable,

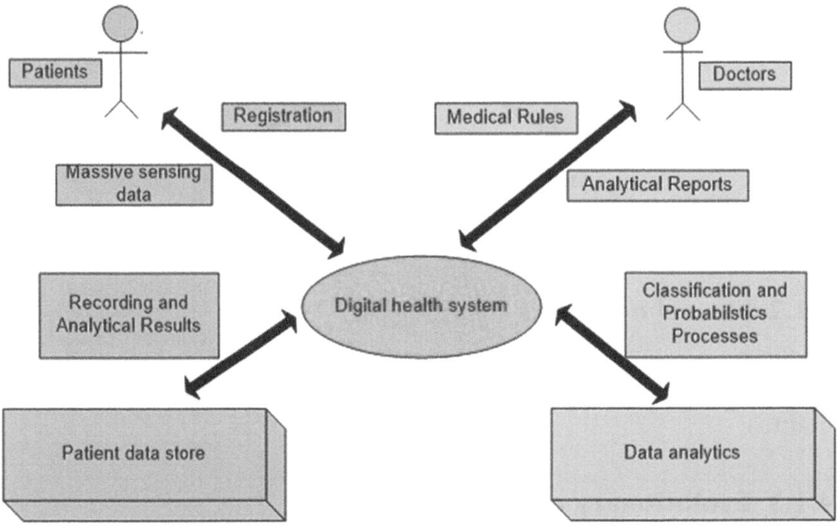

Figure 4.2 Digital health system.

available, transparent, and personalized. It is significant to highlight that digital health systems are expanding globally and embracing a more comprehensive public health approach that guarantees that technological solutions continue to be accessible, safe, and inexpensive for all populations seeking healthcare [5]. Figure 4.2 shows the digital healthcare system.

4.2 COMPUTER INTELLIGENCE APPLICATIONS IN HEALTHCARE

4.2.1 Patient identification

AI-driven facial verification has a lot of promise to improve the workflows in medical facilities. Patient misidentification is a common problem; over 90% of cases are resolved before they cause harm. On the other hand, statistics from 181 U.S. hospitals between 2013 and 2015 showed 7,600 instances of incorrect patient placement, with almost 9% of these mistakes resulting in injury or death [5]. Figure 4.3 presents the computer intelligence in healthcare.

4.2.2 Patient monitoring

Medical personnel utilize computer vision in healthcare to remotely monitor recently discharged patients and outpatients, improving the quality of care. Professionals can effectively monitor patient progress via electronic

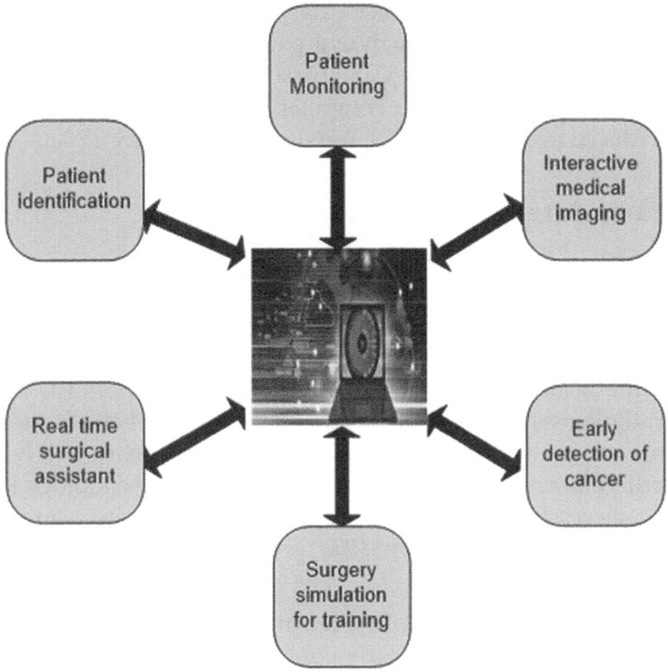

Figure 4.3 Computer intelligence in healthcare.

monitoring [6]. In particular, computer vision is used in fall detection systems in the senior community, making use of sophisticated software and pose estimation methods. The goal of these systems is to lower the number of falls among the elderly, which is a major healthcare concern [7].

4.2.3 Interactive medical imaging

Medical imaging has been transformed by computer vision, allowing for interactive 3D visualization. Medical image processing has been greatly improved by deep learning techniques, opening the door for sophisticated computer vision in healthcare applications [8]. Detailed analyses of dynamic 3D models are made feasible by computer vision, which can help with more accurate medical diagnosis. In contrast to conventional 2D pictures, 3D models offer full insights. For example, state-of-the-art computer vision technology in breast imaging, enable early cancer identification [9].

4.2.4 Real surgical assistant

The field of healthcare is advancing with the introduction of time assistance based on computer vision technology. During surgeries, surgeons benefit from guidance and recommendations through the utilization of reality and advanced algorithms that enhance safety and accuracy. Appreciations, to this technology we can now analyze a patient's structure, in real time and provide more accurate diagnoses and personalized treatments [10]. A recent study found that using real-time computer vision to assist with surgical procedures reduced mistakes by 37% and operation time by 27%.

4.2.5 Early detection of cancer

It is essential to discover brain tumors early in order to lessen their detrimental effects. Innovative computer vision tools have been developed to detect tumors early and treat them quickly, including MASK R-CNN. Deep learning and machine learning techniques are used to train computer vision in the healthcare industry [11]. These systems perform better than human radiologists in terms of speed and accuracy when diagnosing lung and breast cancer by analyzing data from both malignant and healthy tissues. Furthermore, creative apps that make use of AI technology have been incredibly successful; some of these apps are able to detect up to 95% of skin malignancies with accuracy [12].

4.2.6 Surgery simulation for training

Within the medical field, machine vision is useful for teaching surgeons. Pre-emptive surgical knowledge is made possible by this technology, which lowers risks and increases success rates. Cutting-edge apps like Touch Vision provide aspiring surgeons a tactile experience by simulating surgeries realistically using computer vision [13]. By integrating intricate 3D imagery, artificial intelligence approaches are utilized to augment the accuracy of these simulations. This integration of 3D imaging, AI, and machine vision not only improves surgical skills but also raises the competency of aspiring surgeons by providing a realistic, hands-on training environment [14].

4.3 COMPUTER INTELLIGENCE'S ADVANTAGES IN HEALTHCARE

4.3.1 Customized patient care

The computer intelligence makes the quality of patient care much better by analyzing datasets from the patients through which well-personalized insights and advice are generated that leads to improving the outcomes and satisfaction of the patients based on customized interventions and therapies [15]. Figure 4.4 presents the benefits of computer intelligence in healthcare.

4.3.2 Improved analytic speed & accuracy

In industries and professional fields, computer algorithms have proved that they can diagnose problems faster and with more accuracy than human

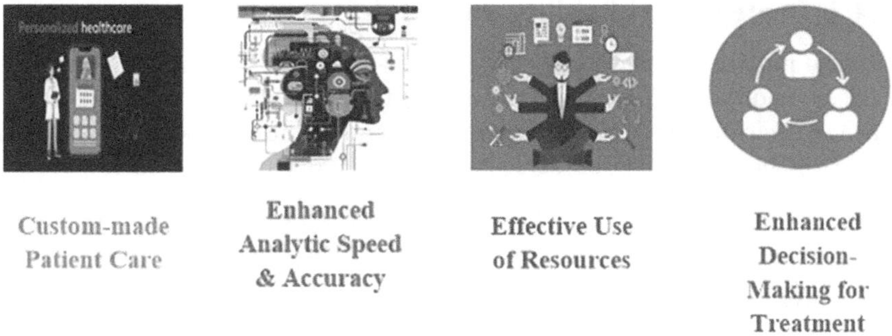

Figure 4.4 Benefits of CI in healthcare.

experts. The integration of computer wisdom in processes minimizes the risks of errors, thus resulting in quicker and more accurate diagnoses [16].

4.3.3 Effective use of resources

Predictive analysis and remote monitoring of patients are examples of CI applications that help optimize the use of resources through efficient treatment and decreased hospital readmission rates, thus saving money and providing better healthcare services [17].

4.3.4 Improved treatment decision making

Personalized treatment plans for each individual and use of their information in decision making and treatment optimization computer intelligence improves client satisfaction and outcomes of treatment [18].

4.4 CHALLENGES

4.4.1 Principled attentions

The use of computer intelligence in healthcare raises ethical questions pertaining to patient autonomy, privacy, transparency, and trust. Accountable and ethical integration of CI in the medical arena would require the upholding of ethical standards and tackling the above issues [19].

4.4.2 Dearth of awareness on computer intelligence

In healthcare, developing trust with patients is essential, and this is particularly true when it comes to computer intelligence-based technologies. Physicians and policymakers recognize that, given the long-standing value of in-person interactions with physicians, the public may be reluctant to rely solely on AI-generated recommendations [20].

4.4.3 Professional liability and healthcare providers

Clinical judgment has traditionally been the purview of qualified and experienced medical professionals. However, the growing use of computer intelligence decision support systems to help with clinical tasks may have an effect on health care providers' professional obligations to their patients [21–22].

4.5 PROSPECTS & FUTURE PATHS

4.5.1 Cooperation Attempts

To promote the creation and uptake of AI solutions in healthcare, cooperation between legislators, AI developers, and healthcare professionals is essential. Multidisciplinary partnerships maximize AI's potential to improve patient outcomes, guarantee ethical practices, and promote a deeper understanding of the healthcare context.

4.5.2 AI that can be explained

Improving explainability of AI algorithms would help foster mutual respect and understanding between patients and medical professionals. Efforts toward developing interpretable AI models and framework will accelerate the adoption of AI in patient-specific care.

4.5.3 Governance and ethical standards

The expansion, application, and assessment of Artificial Intelligence in health care must be guided by strong ethical standards and governance frameworks. These rules should cover accountability, transparency, equity, and responsible use of Artificial Intelligence technologies.

4.5.4 Artificial intelligence in medical education

Artificial Intelligence education should be incorporated into the curricula of medical students, which will enable them to gain acquaintance and assistances that are needed to use AI technologies appropriately. AI literacy, ethical issues, and prudent application of AI in patient care should be the major subjects of education and training programs.

4.6 CONCLUSION

By increasing diagnostic precision and reducing the demand on finite healthcare resources, computer intelligence (CI) has significantly benefited the healthcare sector. It has enhanced the understanding and care for a number of disorders related to the heart, kidneys, and mind. Advanced algorithms analyze vast amounts of patient data, allowing for earlier detection and more accurate diagnosis of these conditions. This leads to improved treatment outcomes and better resource allocation within healthcare facilities.

However, the integration of CI in healthcare comes with concerns about data security and privacy. As cyberattacks have increasingly targeted healthcare systems, incidents of revealed private patient data have raised alarms. These breaches can not only violate patient trust but also potentially harm individuals if their medical information is manipulated or improperly accessed. Additionally, data disruption and shifting can lead to incorrect diagnoses for patients. If data is lost or altered, the nuances that inform a clinician's decision-making process may be compromised. It is crucial for healthcare organizations to implement robust cybersecurity measures to protect sensitive information while leveraging the benefits of computer intelligence. Ensuring data integrity and security will be fundamental in maximizing the advantages of CI in healthcare while safeguarding patients' rights and welfare. Even though AI presents certain challenges in the health industry, it can help patients and enhance industrial efficiency.

REFERENCES

[1] Bajwa, J., Munir, U., Nori, A., & Williams, B. (2021). Artificial intelligence in healthcare: Transforming the practice of medicine. *Future Healthcare Journal*, 8(2), 188–194.

[2] Ansari, M., Dash, B., Sharma, P., & Yathiraju, N. (2022). The impact and limitations of artificial intelligence in cybersecurity: A literature review. *International Journal of Advanced Research in Computer and Communication Engineering*, 11(9), 81–90.

[3] Abishev, O., Balloer, R., Dagan, N., Fahy, N., Hamilton, C., Joshi, I., ... Morley, J. (2019). Digital Health Sytems. *Euro Health*, 25(2), 1–32.

[4] Alahmari, N., Alswedani, S., Alzahrani, A., Katib, I., Albeshri, A., & Mehmood, R. (2022). Musawah: A data-driven AI approach and tool to co-create healthcare services with a case study on cancer disease in Saudi Arabia. *Sustainability*, 14(6), 3313.

[5] Mitchell, M., & Kan, L. (2019). Digital technology and the future of health systems. *Health Systems & Reform*, 5(2), 1–8.

[6] Dash, B., Ansari, M. F., Sharma, P., & Ali, A. (2022). Threats and opportunities with AI-based cyber security intrusion detection: A review. *International Journal of Software Engineering & Applications*, 13(5), 13–21.

[7] Ansari, M. F., Sharma, P. K., & Dash, B. (2022). Prevention of phishing attacks using AI-based cybersecurity awareness training. *International Journal of Smart Sensor and Adhoc Network*, 3(3), 61–72.

[8] Parasuraman, R. (2021). Adaptive automation: From theory to practice. *Proceedings of the Human Factors and Ergonomics Society Annual Meeting*, 1(1), 25–37.

[9] Dash, B. (2021). A hybrid solution for extracting information from unstructured data using optical character recognition (OCR) with natural language processing (NLP).
[10] Mostafa, F. A., Elrefaei, L.A., Fouda, M. M., & Hossam, A. (2022). A survey on AI techniques for thoracic diseases diagnosis using medical images. *Diagnostics (Basel)*, 12, 3034. https://doi.org/10.3390/diagnostics12123034
[11] Olive-Gadea, M., Crespo, C., Granes, C., et al. (2020). Deep learning-based software to identify large vessel occlusion on non-contrast computed tomography. *Stroke*, 51, 3133–3137. https://doi.org/10.1161/STROKEAHA.120.030326
[12] Lin, K., Liu, J., & Gao, J. (2022). AI-driven decision making for auxiliary diagnosis of epidemic diseases. *IEEE Transactions on Molecular, Biological, and Multi-Scale Communications*, 8, 9–16. https://doi.org/10.1109/TMBMC.2021.3120646
[13] Comito, C., Falcone, D., & Forestiero, A. (2022). AI-driven clinical decision support: enhancing disease diagnosis exploiting patient's similarity. *IEEE Access*, 10, 6878–6888. https://doi.org/10.1109/ACCESS.2022.3142100
[14] Iqbal, J., Jahangir, K., Mashkoor, Y., et al. (2022). The future of artificial intelligence in neurosurgery: A narrative review. *Surgical Neurology International*, 13, 536. https://doi.org/10.25259/SNI_877_2022
[15] Chen, P., Chen Xu, R., & Chen, N., et al. (2021). Detection of metastatic tumor cells in the bone marrow aspirate smears by artificial intelligence (AI)-based Morphogo system. *Frontiers in Oncology*, 11, 742395. https://doi.org/10.3389/fonc.2021.742395
[16] Bajwa, J., Munir, U., Nori, A., & Williams, B. (2021). Artificial intelligence in healthcare: transforming the practice of medicine. *Future Healthcare Journal*, 8(2), e188–e194. https://doi.org/10.7861/fhj.2021-0095
[17] Johnson, K. B., Wei, W. Q., Weeraratne, D., Frisse, M. E., Misulis, K., Rhee, K., Zhao, J., & Snowdon, J. L. (2021). Precision medicine, AI, and the future of personalized health care. *Clinical and Translational Science*, 14(1), 86–93. https://doi.org/10.1111/cts.12884
[18] Topol, E. J. (2019). High-performance medicine: The convergence of human and artificial intelligence. *Nature Medicine*, 25(1), 44–56.
[19] Liu, X., Faes, L., Kale, A. U., Wagner, S. K., Fu, D. J., Bruynseels, A., ... & De Fauw, J. (2020). A comparison of deep learning performance against health-care professionals in detecting diseases from medical imaging: a systematic review and meta-analysis. *The Lancet Digital Health*, 2(6), e271–e297.
[20] Char, D. S., Shah, N. H., Magnus, D., & Hripcsak, G. (2020). Implementing machine learning in health care—Addressing ethical challenges. *New England Journal of Medicine*, 382(11), 981–983.
[21] Sun, T. Q., & Medaglia, R. (2019). Mapping the challenges of artificial intelligence on the public sector: evidence from public healthcare. *Government Information Quarterly*, 36(2), 368–383. https://doi.org/10.1016/j.giq.2018.09.008
[22] Carter, S. M., Rogers, W., Win, K. T., Frazer, H., Richards, B., & Houssami, N. (2020). The ethical, legal and social implications of using artificial intelligence systems in breast cancer care. *Breast*, 49, 25–32. https://doi.org/10.1016/j.breast.2019.10.001

Chapter 5

Harnessing the potential of reinforcement learning for effective decision strategies

Gaurav Pandey, Devendra Rawat, and Amit Kukker

5.1 INTRODUCTION: DECISION-MAKING AND REINFORCEMENT LEARNING

AI enabled decision-making is one of the most significant uses of AI these days. Decision-making is the avenue that provides the optimal course of action when considering all the possibilities of uncertainty. Decision-making in AI implements intelligent methods such as robots capable of autonomous decisions in a similar way to humans. Decision-making is crucial and of utmost importance. It considers the changing conditions and complicated environments, and thus decisions are adaptive in nature.

5.2 IMPORTANCE OF DECISION-MAKING IN REAL-WORLD APPLICATIONS

Across all the fields and sectors, AI provides decision making, some of the popular examples are listed below:

i. **Autonomous Vehicles**:
Autonomous vehicles are a recent technological advance that requires real-time decision-making for self-driving automobiles, to navigate, obstacle avoidance, and to select safe-driving preferences. By self-driving cars the constant monitoring of the surroundings and accordingly adaptive decisions are required to ensure the safety of occupants and other road users [1].

ii. **Robotics**:
Decision-making in robotics allows robotic systems to perform complicated tasks precisely such as delivery, inspection, and exploration [2, 3]. The robots need to have accurate information about their surroundings so that it can avoid impediments and can perform the job intended effectively.

DOI: 10.1201/9781003568995-5

iii. **Healthcare:**
Use of AI in healthcare is increasing tremendously these days. These systems help professionals to diagnose diseases by analyzing medical-data patient history, health records, and disease symptoms [3, 4]. Pharmaceutical applications have hugely enabled decision-making in drug discovery, delivery, and personalized medication.

iv. **Finance:**
The finance sector has huge data-driven decision making in various applications such as stock market trading. AI algorithms monitor market trends, detect patterns, and execute real-time trade choices to maximize profits or reduce risks. Another important aspect of AI systems is to utilize and estimate the loans, and manage risk possibilities [5, 1].

v. **Energy and Environment**
AI systems provide impactful decisions in the energy and environmental sector, these include optimum utilization of renewable resources, controlling energy requirements from grids and ensuring efficient distribution. In environmental studies, algorithms for decision-making help to evaluate climate prediction, soil parameters, and the impact of numerous environmental policies and activities [6].

5.3 FUNDAMENTALS OF REINFORCEMENT LEARNING

Reinforcement Learning (RL) is a category of machine-learning algorithms in which an autonomous agent engages with its environment, acquiring information about its present state and makes decisions utilizing this knowledge. The agent tends to select actions that maximizes cumulative reward as it explores the environment. The environment delivers a reward signal (either positive or negative). The aim of the agent is to maximize the predicted cumulative reward signal throughout interactions [7]. RL-framework is depicted in Figure 5.1.

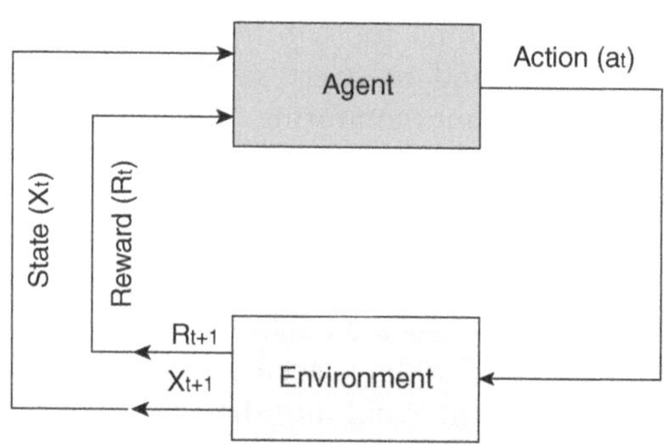

Figure 5.1 Representation of the RL framework.

Policy is the behavior of an autonomous agent in an environment. Policy takes current state as input and produces action. A policy can be categorized as either deterministic or stochastic.

Agent: In RL, it is a decision-making unit that interacts with the environment to adopt and maximize rewards through its actions.
Environment: The physical reality in which the agent functions.
State: It describes the agent's current status or condition within the environment.
Action: In RL, the agent takes an action, which is a move in reaction to its existing state within the environment.
Reward: It is the signal or feedback generated by the environment in reaction to the action of the agent.

5.4 TYPES OF REINFORCEMENT LEARNING

There are different approaches to RL implementation which are presented below.

i. **Model-Based vs Model-Free RL Algorithms**
Model-based reinforcement learning (MBRL) entails building an explicit model of the environment that determines state transitions and rewards, similar to the concept of a full Markov Decision Process (MDP). This information is frequently shown as a mathematical model that details the probabilities, rewards, and state changes connected to every action. Alternatively expressed, MBRL methods use an internal model of the environment to simulate action outcomes and calculate action values. By capturing the states, transition probabilities, and rewards of the environment, this model enables the agent to plan and evaluate actions effectively. The agent can create a model of the environment through experiential learning, enabling it to explore multiple map routes and decide on the optimal course of action [3].
On the other hand, model-free methods directly determine the optimal course of action without explicitly modeling environmental dynamics like the reward function and transition probabilities. In other words, model-free reinforcement learning is a decision-making process where the value of various actions is determined by trial-and-error interaction with the black box when there is no world model. This approach is more data-driven and less dependent on prior environmental knowledge [3]. To direct the decision of action, the predicted values are updated trial-by-trial through the computation of prediction errors. After learning converges, the optimal technique for

choosing an action is model-free reinforcement learning. However, this technique is less adaptable to rapid changes in the environment because the values are based on accumulated previous experience, and a lot of trial-and-error experiences are required to get it more accurate. Q-learning and SARSA are some popular model-free reinforcement learning products.

ii. **Value-Based and Policy-Based RL Algorithms**

The agent generates a value function in value-based RL that assigns value to every state-action pair. This approach is effective when there is a finite set of possible actions. Value-based approaches focus on determining the numerical worth of being in a specific state or taking a particular action, typically measured by expected rewards [3]. A primary objective of these methods is to identify the optimal state-value or action-value function, which can then be used to derive the policy. This process is often referred to as an indirect method of policy generation, as the policy is obtained by first-learning value functions.

In Policy-based RL the agent directly learns a policy, which maps states to actions, aiming to increase cumulative rewards over time. In this, the agent focuses on finding optimal policy rather than computing value function. It is useful when action space is stochastic or continuous [3]. Policy-based approaches have demonstrated success in learning stochastic policies, which are particularly effective in environments with continuous or high-dimensional action spaces. This capability often makes them more advantageous than deterministic approaches, especially in scenarios requiring exploratory or probabilistic actions. Example- Examples*: REINFORCE, Trust Region Policy Optimization (TRPO), Proximal Policy Optimization (PPO).*

iii. **On-Policy and Off-Policy Based RL Algorithms**

An on-policy algorithm relies on a single policy, which the agent uses to select actions in a given state x with the objective of maximizing cumulative rewards. Off-policy algorithms utilize two distinct policies: the behavior policy and the target policy. In target policy, the agent aims to optimize and learn over time. In contrast, the behavior policy generates the data or experiences that are used during the learning process. This separation allows the agent to learn from data generated by different policies or even from pre-existing datasets, enhancing flexibility and efficiency. Since every off-policy approach is convertible made in an on-policy algorithm by adjusting the target policy to match the behavior policy, off-policy learning is an extension of on-policy learning [3]. Figure 5.2 depicts the classification flowchart for RL algorithms.

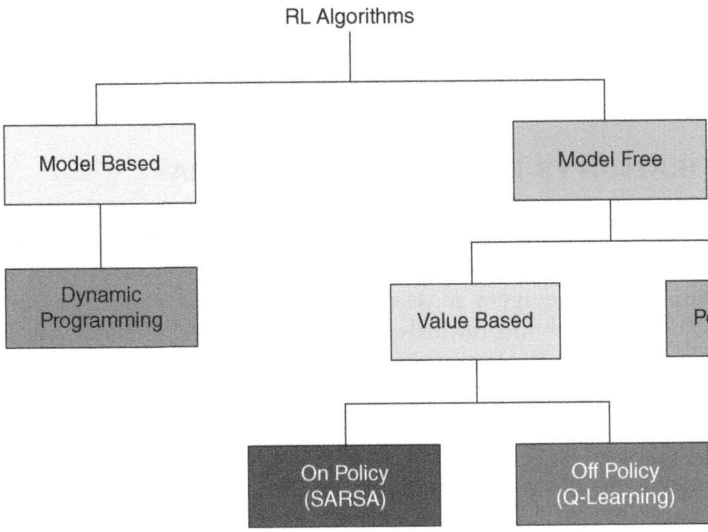

Figure 5.2 Classification of RL algorithms.

5.5 THE MARKOV DECISION PROCESS (MDP) FRAMEWORK

The mathematical framework that simulates the interaction of an agent with its environment is termed as MDP. This aims to make decisions and maximize rewards over time. It relies on the Markov property, which asserts that the next state is impacted solely by the current state and action, ignoring any previous states or actions [3, 7]. Simply said, it focuses on forecasting the "next outcome" based purely on the existing situation. MDP is defined by the fallowing 5-tuple.

a) **State (X)**: It refers to the numerous environmental circumstances or scenarios.
b) **Actions (A)**: The collection of all potential actions the agent can perform.
c) **Transition Probability (P)**: The transition function $p(x', r \mid x, a)$ determines the likelihood of transitioning to a new state (x') and getting a reward r while the agent is in the current state and takes action. In essence, it records how the environment reacts to the agent's activities by evaluating the possibilities of the resultant state and reward.
d) **Reward (R)**: The response received by the agent after performing an action in a particular state, representing the immediate reward or benefit of that action.

e) **Discount Factor** (γ): A value between 0 and 1 that represents the importance of future rewards. A lower value favors immediate rewards, while a higher value gives more weight to long-term rewards.

5.6 EXPLORATION VS. EXPLOITATION DILEMMA

To maximize rewards, an RL agent should prefer actions that have been previously attempted and have proven beneficial in the past. However, to identify such actions, the agent must experiment with choices it has not previously selected. To obtain rewards, the agent must exploit its existing knowledge to gain rewards. However, it must also explore to choose better courses of action in the future [8]. The problem is that focusing primarily on exploration or exploitation will result in failure to complete the task. Both must be balanced for success. The agent must experiment with various actions and gradually give preference to those that seem to work best.

5.7 THE REWARD SIGNAL AND ITS ROLE IN LEARNING

In an RL task, the reward signal defines the objective. The environment assigns a single numerical value, known as the reward, to the agent at every time step. The agent's main goal is to maximize the overall reward in the long run. The reward signal determines which occurrences are positive or negative for the agent. The reward signal is the fundamental reason for changing the policy. If an action chosen by the policy results in a poor reward, the policy may be modified to select a different action in similar instances in the future. The reward signal conveys what is good in the current moment whereas, a value function explains whatever is good in the prolonged term. Rewards are primary, while values are secondary as they forecast future rewards [8]. Values cannot exist without rewards, and evaluating values is done solely for the purpose of maximizing rewards. The environment provides rewards directly, but values must be assessed based on an agent's lifetime observations.

5.8 CORE ALGORITHMS IN REINFORCEMENT LEARNING

5.8.1 Temporal difference (TD) Learning

In RL temporal difference learning is a fundamental approach for the evaluation and improvement of policies based on the idea of learning directly from raw experience, without exact knowledge of the model of the environment. It combines ideas from both the Monte Carlo method and dynamic programming. In TD learning the agents updates the value estimates of states based on the difference between two successive estimates, rather than waiting for a complete episode to end [3]. The temporal part refers to using future predictions to update current predictions, while the difference refers to the discrepancy between the expected and the actual value observed. At each time step Temporal Difference (TD) approaches incrementally upgrade their value function utilizing the reward seen at that step and the estimated value of the subsequent. This specific technique is called TD (0). This equation shows the mathematical representation of how the value function or action-value function is updated in TD (0) methods [8, 9]:

$$V(x_t) \leftarrow V(x_t) + \alpha\left[R_{t+1} + \gamma V(x_{t+1}) - V(x_t)\right] \tag{5.1}$$

5.8.1.1 TD error

The main idea in TD learning is the TD error, defined as:

$$\delta_t = R_t + \lambda . V(x_{t+1}) - V(x_t) \tag{5.2}$$

δ_t = Temporal difference error at time t.
R_t = Reward obtained after transition from state s_t to s_{t+1}.
γ = Discount factor
$V(x_t)$ = Value estimate of the current state s_t.
$V(x_{t+1})$ = Value estimate of the next state s_{t+1}.

5.8.1.2 TD learning update rule

The TD learning update control adjusts the value of $V(x_t)$ based on the TD error:

$$V(x_t) \leftarrow V(x_t) + \alpha\delta_t \tag{5.3}$$

Q-learning and SARSA learning are two TD control algorithms.

The optimal policy (π) selects the action (a) that maximizes the action-value function.

$$\pi(x) = \arg \underset{a}{max}\ q_\pi(x,a) \tag{5.4}$$

5.8.2 Q-learning

Watkins proposed Q-learning as a **model-free** optimization strategy for decision-making in uncertain systems. It frames the problem of sequential decision-making as a ***Markov Decision Process*** (**MDP**, through agent interactions within the environment to maximize cumulative rewards over time [10]. Q-learning is a TD technique of learning in which the agent is not required to learn the environment's transition model [8]. Instead, it emphasizes learning the Q-value function. It is also an off-policy approach in RL that approximates the optimal Q-value [9]. The Q-value, also referred to as the **state-action value**, represents the expected cumulative reward for taking a specific action a in a given state x and then following the optimal policy. The Q-value function is commonly denoted as $Q(x,a)$. The Q-values are updated using the following formula:

$$Q(x,a) \leftarrow Q(x,a) + \alpha\left(R + \gamma \; max_{a'} Q(x',a') - Q(x,a)\right) \tag{5.5}$$

where x and x' are current state and next state, respectively. a and a' are current action and next action, respectively. γ is the **discount factor**, which determines the importance of future rewards, while α, ranging from 0 to 1, is the **learning rate** that controls the speed at which the Q-value is adjusted based on the newly received reward.

Arbitrarily Initialization of $Q(x,a)$
Repeat (for each episode)
Initialize x
Repeat until x is terminal (for every episode stage):
Choose A from X using policy derived from Q (e.g., ε-greedy)
Take action a, observe reward R, state x'

$$Q(x,a) \leftarrow Q(x,a) + \alpha\left(R + \gamma \; max_{a'} Q(x',a') - Q(x,a)\right) \tag{5.6}$$

$$x \leftarrow x', a \leftarrow a';$$

Until x is terminal.

5.8.3 SARSA-learning

SARSA utilizes an on-policy TD update approach for learning. In SARSA, the updated state-action value is influenced by the received reward, the current state's value, and the value of the next chosen state-action pair [11]. In this algorithm, The Q-values are updated using the following formula [8, 9]:

$$Q(x,a) \leftarrow Q(x,a) + \alpha\left(R + \gamma . Q(x',a') - Q(x,a)\right) \tag{5.7}$$

Comparison of Q-Learning and SARSA Algorithms is shown in Figure 5.3.

Arbitrarily Initialization of $Q(x,a)$
Repeat (for each episode)
Initialize x
Choose A from X using policy derived from Q (e.g., ε -greedy)
Repeat until x is terminal (for every episode stage):
Take action a, observe reward R, state x'

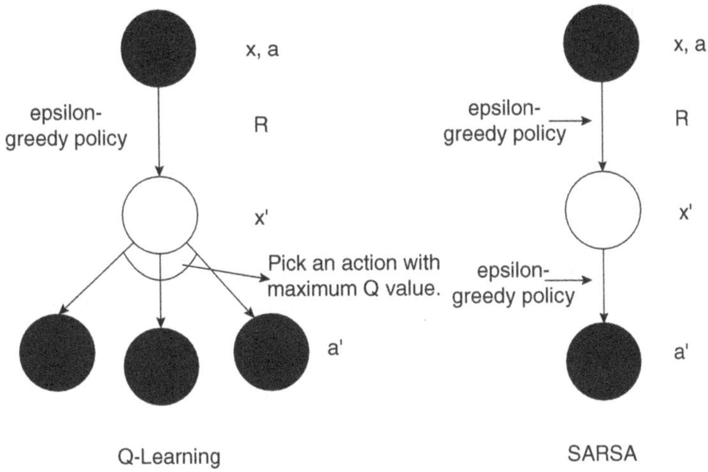

Figure 5.3 Comparison of Q-learning and SARSA algorithms.

Choose a 'from s' using policy derived from Q (e.g. -greedy)

$$Q(x,a) \leftarrow Q(x,a) + \alpha\left(R + \gamma.Q(x',a') - Q(x,a)\right) \tag{5.8}$$

$$x \leftarrow x', a \leftarrow a'$$

5.8.4 Deep-Q-learning

Q learning combined with deep learning is called deep Q learning. Deep-learning research has enabled machines to solve complex decision-making problems that were previously considered unfeasible. Q-Learning generates a Q-table containing values indicating the quality of state-action pairs. The table has dimensions (s,a), with *s* representing the number of possible states and *a* representing the number of accessible actions. Essentially, the Q-table assigns a Q-value to each state-action pair, which represents the expected future reward. However, in a real-life situation, the number of states could be enormous, making it computationally impossible to construct a table. In order to overcome this limitation, we employ a Q-function instead of a Q-table, which accomplishes the identical goal of mapping action and state pairings to a Q value. Because neural networks are adept at modeling complex functions, we may estimate this Q function using a neural network known as a Deep Q Network [11]. The DQN framework consists of two neural models: the Q-network and the Target network, as well as an experience replay buffer. The Q-network, which serves as the agent's representation in Deep Q-Learning, is trained to approximate the optimal state-action value function. The Q-network accepts the current state as input and returns the Q value $(Q(x,a))$ for each conceivable action. Target network is a clone of Q-network that is updated less regularly. The target network is employed to stabilize training and limit the possibility of divergence. After interaction with environment the agent is collecting information in the form of state, action, reward, and next tuples to improve the experience. Replay buffer stores every experience, which samples a random batch of experiences and updates the Q-values, that cares for the stability of training. For every experience (x, a, R, x') within the batch, the target Q-value is calculated using the bellman equation [12, 13]. The Q-value prediction $Q(x, a)$ is compared with the Q-value, and a loss is calculated which is used to update the network using gradient descent accordingly. The DQN's aim is to evaluate the best policy by choosing actions which results in best expected reward for every given state.

Consider the state space $X = \{x_1, x_2, x_3\}$, denoting all potential states that an agent may encounter. The action space $A = \{a_1, a_2\}$ encompasses all conceivable actions the agent performs to interact with the environment. The replay buffer, with a capacity of four, can retain up to four experiences at any given time, enabling the agent to sample and learn from past interactions. The following steps outline the usual Deep Q-Network (DQN) algorithm procedure.

5.8.5 Initialization

The replay buffer is initially empty, and the Q-network is seeded with random weights for estimating Q-values. The target network is originally a replica of the Q-network, and its goal is to offer steady targets throughout the training phase.

5.8.6 Collect experiences

Starting at state x, the agent takes an action a, receives a reward R (such as 1), and transitions to a new state x' to continue interacting with the environment. This experience, represented as the tuple $(x_1, a_1, 1, x_2)$ is subsequently stored in the buffer.

5.8.7 Subsequent interaction

More experiences are gathered as the agent keeps interacting with the environment. The replay buffer uses a FIFO (First In, First Out) approach, whereby new experiences replace the oldest ones after it reaches its maximum of four experiences.

5.8.8 Selecting a batch

The replay buffer is used to select a random set of experiences. Target Q-values are calculated and the Q-network is trained utilizing this batch.

5.8.9 Determine target Q-values

For each experience tuple (a, R, x')in the batch target Q-value is:

$$TargetQ_{value} = R + \gamma . \max_{a'} Q_{target}(x', a') \tag{5.9}$$

5.8.10 Update the Q-network

The weights of the Q-network are adjusted through backpropagation, utilizing the Mean Squared Error (MSE) loss function for the update.

$$Loss = \frac{1}{K} \sum_{i=1}^{K} \left(predictedQ_{value} - TargetQ_{value}\right)^2 \tag{5.10}$$

5.8.11 Update the target network

After a predetermined number of steps, update the target network to match the Q-network.

$$\theta_{target} \leftarrow \theta_{Q-network} \tag{5.11}$$

In the DQN algorithm, the aforementioned procedure represents one update cycle. The agent refines its policy and learns the best course of action for optimum rewards by repeating these cycles. The process flow of DQN algorithm is depicted in Figure 5.4.

5.8.12 Policy gradient methods

The policy gradient approach is a reinforcement learning strategy in which the agent concentrates on improving its behavior to maximize total reward

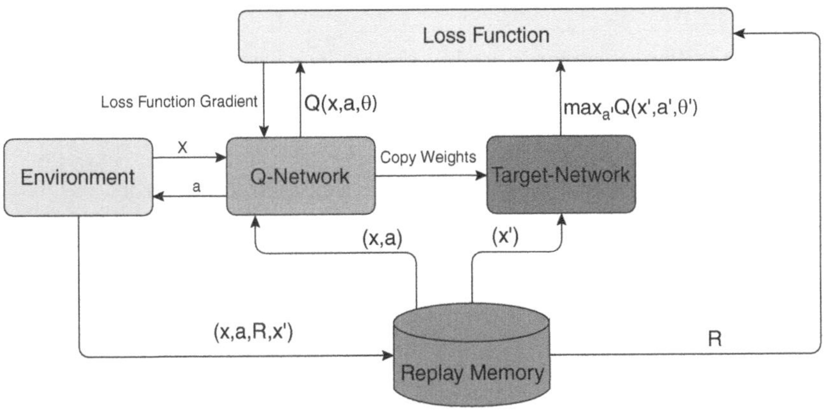

Figure 5.4 Process-flow of DQN algorithm.

by directly changing and refining the policy [8]. The policy is written as $\pi_\beta(x,a;\beta)$, indicating the likelihood of picking action a in state x. The vector of parameters that define the policy is denoted as β. The policy parameters are updated using the following update rule:

$$\beta = \beta + \alpha\nabla_\beta J(\beta) \tag{5.12}$$

Where $J(\bar{\beta})$ is the reward function. $\nabla_{\bar{\beta}} J(\bar{\beta})$ represents the gradient of this reward function with respect to the policy parameters β, which tells us how to adjust β to increase the expected reward and $\alpha$is the learning rate. The gradient of the reward function can be expressed by following equation.

$$\nabla_\beta J(\beta) = E_{\pi_\beta}\left[\nabla_\beta \log \pi_\beta(x,a;\beta) Q^\pi(x,a)\right] \tag{5.13}$$

$Q^\pi(x,a)$ is the **action-value function** under the policy π_β [13].

5.8.13 Actor-critic algorithms

Actor-critic techniques are a popular reinforcement learning framework that relies on two distinct components working together: the actor, which determines the actions to take, and the critic, which evaluates those actions. The actor learns a policy $\pi_\beta()$ which optimizes the expected reward, whereas the critic's goal is to learn an accurate value function that evaluates actions of actor. By combining the policy gradient, that guides the actor in improving its policy, with the value function used by critics to evaluate the actions, the Actor-Critic technique formulates its objective function [11]. One of the advantages of actor-critical approaches is that it reduces variance during policy updates. The Critic acts as a baseline, hence stabilizing the learning process; Actor updates parameters (β) using policy gradient [8, 9].

$$\beta = \beta + \alpha\nabla_\beta J(\beta) \tag{5.14}$$

Temporal difference error is used to update its parameter (χ) by critic.

$$\delta_t = r_t + \gamma V_\chi(x_{t+1}) - V_\chi(x_t) \tag{5.15}$$

The Critic uses gradient descent to reduce the squared TD error:

$$\chi \leftarrow \chi - \alpha.\delta.\nabla_\chi V_\chi(x) \tag{5.16}$$

The Actor employs the TD error δ and the policy gradient concept.

$$\nabla_\beta J(\beta) = E_{\pi_\beta}\left[\nabla_\beta \log \pi_\beta(x,a;\beta).\delta\right] \tag{5.17}$$

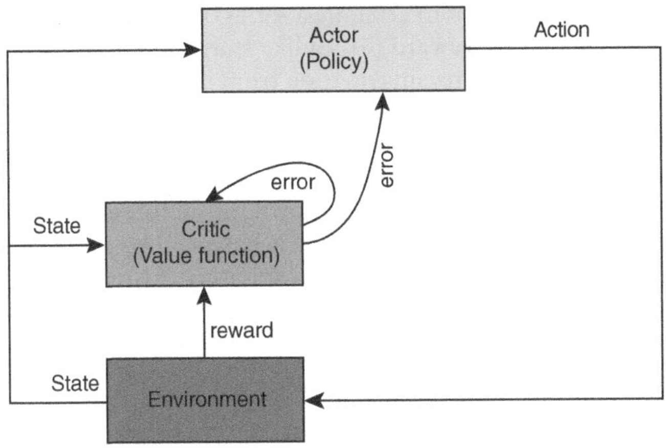

Figure 5.5 Actor-critic algorithm.

Critic updates the value function parameters *w* using TD error. Figure 5.5 describes the Actor-Critic algorithm.

5.9 REINFORCEMENT LEARNING IN DECISION-MAKING FRAMEWORKS

Reinforcement Learning (RL) is a type of machine learning that aims to tackle sequential decision-making issues, especially with complex and dynamic situations. RL promotes optimal decision-making by allowing an agent to gain knowledge from interactions with the environment and refine its approaches over time. Reinforcement learning (RL) has emerged as a useful tool for addressing difficult decision-making issues in a variety of fields, including education, healthcare, and population research. The core principle behind RL is that an agent, trains to make decisions by interacting with its environment, evaluating the results of its actions, and constantly updating its decision-making policy in order to maximize its long-term rewards.

5.10 HOW RL AIDS COMPLEX DECISION-MAKING

Here's an overview of how RL helps with complex decision-making:

a) **Making Sequential Decisions**
In many real-world circumstances, decisions are made sequentially, with the outcome of one selection influencing subsequent options. RL excels in such environments by evaluating the long-term implications

of actions, allowing it to generate a series of decisions that result in the biggest cumulative reward [8]. Unlike static optimization approaches, RL constantly acquires information from its surroundings and adjusts its strategy in response to new inputs.

b) **Addressing Uncertainty**
Complex decision-making frequently entails significant uncertainty, with results influenced by unknown or unforeseen circumstances. RL addresses this through striking a balance between exploring new actions to collect information and exploiting existing actions for maximum rewards. The agent modifies its policy whenever it finds new states or unexpected outcomes, making it adaptable to changing environments [8].

c) **Maximizing Long-Term Reward**
Many difficult challenges necessitate weighing long-term benefits against short-term rewards. RL naturally incorporates this by optimizing cumulative rewards rather than concentrating on short-term results [8].

d) **Multiple-Agent Systems**
Many complicated systems involve several decision-makers. RL facilitates multi-agent settings by allowing agents to develop cooperative or competing strategies [14].

e) **The ability to scale to high-dimensional issues**
Complex decision-making frequently requires managing several variables and huge state spaces. RL scales effectively in such settings by utilizing approximation techniques. Advanced approaches, such as Deep RL, employ neural networks to estimate value functions and policies in highly dimensional environments.

5.11 MULTI-AGENT REINFORCEMENT LEARNING (MARL)

Multi-agent systems have applications in traffic control, network packet routing, energy distribution, and robotic coordination, among others. In multi-agent environments, single-agent reinforcement learning (RL) techniques typically fail to achieve optimal outcomes. This is because the environment becomes dynamic and unpredictable from each agent's perspective. The environment's behavior changes as the outcomes or rewards of one agent's actions depend on the actions taken by other agents. This phenomenon, known as environmental non-stationarity, represents the primary challenge in developing effective multi-agent RL (MARL) systems [15, 16].

A Markov Decision Process (MDP) is not sufficient to describe the environment when multiple agents are involved because the actions of one agent influence the state transitions for all agents. Instead, we use Markov games (MGs), also known as stochastic games, which are an extension of MDPs designed for multi-agent environments.

A tuple $\left(K, XK, X, \{A^j\}_{j\in K}, f, \alpha\right)$ is the mathematical representation of a MGs where

1. $K = 1,2,3,\ldots,K$ is the set of $K > 1$ agents
2. Xdenotes the collection of all potential environmental states.
3. A^j are the Action Space of Agent $j$and A: $A^1 \times A^2 \times A^3 \times \ldots . \times A^K$is called collective action space.
4. $f : X \times A \rightarrow [0,1]$ is the transition probability function.
5. $O(x', a, o) \geq 0$
   $\sum_{o\in\Omega} O(x', a, o) = 1$ is the reward functions that represents the instantaneous reward earned by the j-th agent, transitioning from (x,a)tox'.
6. $\alpha \in [0,1]$ is the discount factor.

5.12 PARTIALLY OBSERVABLE MARKOV DECISION PROCESS (POMDPS)

A partially observable Markov decision process (POMDP) enables optimal decision-making in situations where the agent can only see part of the environment, unlike the MDP model, which assumes the agent has full visibility of the environment. In the **POMDP** model, the agent cannot directly observe the full state $x \in X$of the environment at any given time. Instead, it perceives a partial representation of the state as an **observation** $o \in \Omega$. This observation is provided by the observation function $O :: X \times A \times \Omega \rightarrow [0,1]$. When the agent takes an action $a \in A$, the system's state changes stochastically, reaching the new state x'. The agent only receives the observation o', which was generated according to observation function [17, 18]. The agent then receives a reward $r \in R$, much as the MDP. These observations may not always be precise and are dependent on the condition of the environment after the agent has performed an action.

Two conditions must be met for the observation function (O) in a POMDP to be considered a legitimate probability distribution [19]:

a) $O(x', a, o) \geq 0$: The probability of receiving observation $o$after transitioning to state x' and performing action a must always be equal to or greater than zero.

b) $\sum_{o\in\Omega} O(x', a, o) = 1$: The total probability of all possible observations for a specific state x' and action a must equal 1.

5.13 CHALLENGES IN REINFORCEMENT LEARNING FOR DECISION-MAKING

5.13.1 Sample efficiency and scalability

Reinforcement learning (RL) is an effective method for training agents for making decisions in complicated environments. Data efficiency refers to the ability to gain knowledge from minimal amounts of data. However, data efficiency remains a significant challenge. In contrast to supervised learning, which involves learning from labeled datasets, RL agents require interaction with their surroundings to collect information. This procedure often requires a large quantity of data since agents must experiment with different actions, analyze the results, and iteratively refine their tactics. In complex jobs like robotics or gaming, millions of interactions may be required to achieve proficiency. Collecting data in real-life situations such as autonomous vehicles or industrial robotics can be expensive, time-consuming, and occasionally dangerous, therefore sample efficiency is a crucial issue [20, 21]. As the environment or goal becomes more complicated, RL algorithms' computing and data requirements increase considerably. Scaling RL models to accommodate high-dimensional state spaces, complicated dynamics, and systems with several agents can be challenging [22].

5.13.2 Reward Engineering and Its Implications

Reward engineering is the process of designing and defining the reward function in a reinforcement learning (RL) environment. The reward function determines how an agent sees the success or failures of its actions to the environment. In essence, it delivers feedback to the agent to help it learn. It might be difficult to strike the correct balance in reward design to ensure that the agent trains effectively and fits with real-world objectives. Consider training an RL agent to play a video game like Pac-Man. A poorly designed reward function might only provide a positive signal when the agent succeeds, making it harder for the agent to develop effective strategies [22]. In comparison, a well-designed reward function may deliver smaller payouts for eating dots, escaping ghosts, and collecting power-ups, directing the agent through the many components of the game.

5.13.3 Safe exploration in critical environments

Safe exploration is important since it allows an RL agent (such as a robot or self-driving car) to learn by trying different things without letting anything that could hurt or damage them get in the way. While dealing with risky situations, like driving a car in heavy traffic or a robotic surgery, it becomes of the utmost importance that the agent must not engage in unsafe actions that could harm people or damage property. To address this, specific controlled methods are required that allow the agent to explore and learn, that ensures safety and are capable of improving the learning mechanism [22]. For example, simulation exercises (virtual environments) can be framed for agents where they can make mistakes without any real-world consequences under the constraints that limit risky actions while the agent is learning.

5.13.4 Transfer learning in RL

The ability to implement the specific knowledge acquired from one activity or environment to another that is similar, is known as transfer learning in RL. This learning seems essential in RL because the training of the agents from the start in every new environment can be highly complex and time-consuming. Transfer learning improves the learning process by enabling the agents to apply previously learned experiences to new contexts. However, transfer learning in reinforcement learning has a number of challenges, including figuring out how to effectively transfer knowledge across various contexts and making sure the agent doesn't perform poorly as a result of task changes [21]. To ensure that the agent adapts successfully and without unintended consequences, the information transmission mechanism needs to be carefully controlled.

5.14 CONCLUSION

RL methods demonstrate the potential to support intelligent decision-making and offer an optimal robust framework to deal with complex, dynamic, and uncertain challenges. These methods implement agents capable of learning optimal strategies through trial and error and thus provide real-time problem-solving ability. This chapter summarizes the RL methodologies and presents the significant advancements along with implementation and how these methods have impacted numerous sectors.

RL techniques are continuously evolving by innovations in algorithms, enhancements in computational power despite several challenges like computational demands, scalability, and other ethical considerations. The amalgamation of RL with recent developments involving deep learning methods, multi-agent systems, and edge computing has improved the frontiers of RL. Harnessing RL's potential for efficient decision-making techniques not only revolutionizes existing procedures but also paves the way for further advancements in artificial intelligence, enabling systems to make wise, independent, and significant choices.

REFERENCES

[1] J. Ghahremani Nahr, H. Nozari, and M. E. Sadeghi, "Artificial Intelligence and Machine Learning for Real-world Problems (A Survey)," 2021. [Online]. Available: www.ijie.ir, doi: 10.59615/ijie.1.3.38

[2] R. Alterovitz, S. Koenig, and M. Likhachev, "Robot Planning in the Real World: Research Challenges and Opportunities," 2016, doi: 10.1609/aimag.v37i2.2651

[3] M. N. Al-Hamadani, M. A. Fadhel, L. Alzubaidi, and B. Harangi, "Reinforcement Learning Algorithms and Applications in Healthcare and Robotics: A Comprehensive and Systematic Review," *Sensors*, vol. 24, no. 8, p. 2461, 2024, doi: 10.3390/s24082461

[4] M. Wagner et al., "The Importance of Machine Learning in Autonomous Actions for Surgical Decision Making," *Artificial Intelligence Surgery*, vol. 2, no. 2, pp. 64–79, 2022, doi: 10.20517/ais.2022.02

[5] V. Pillai, "Integrating AI-Driven Techniques in Big Data Analytics: Enhancing Decision-Making in Financial Markets," *International Journal of Engineering and Computer Science*, vol. 12, no. 07, pp. 25774–25788, Jul. 2023, doi: 10.18535/ijecs/v12i07.4745

[6] N. M. Kumar et al., "Distributed Energy Resources and the Application of AI, IoT, and Blockchain, in Smart Grids," *MDPI AG*, Nov. 01, 2020, doi: 10.3390/en13215739

[7] D. Abel, J. Macglashan, and M. L. Littman, "Reinforcement Learning as a Framework for Ethical Decision Making." [Online]. Available: www.aaai.org, doi: 10.1016/j.robot.2015.11.012

[8] R. S. Sutton and A. G. Barto, 2018. *Reinforcement Learning: An Introduction*. MIT Press.

[9] C. Szepesvári, 2022. *Algorithms for Reinforcement Learning*. Springer nature.

[10] A. Kukker, R. Sharma, G. Pandey, and M. Faseehuddin, "Fuzzy Lattices Assisted EJAYA Q-Learning for Automated Pulmonary Diseases Classification," *Biomedical Physics & Engineering Express*, vol. 10, no. 6, p. 065001, Nov. 2024, doi: 10.1088/2057-1976/ad72f8

[11] B. Singh, R. Kumar, and V. P. Singh, "Reinforcement Learning in Robotic Applications: A Comprehensive Survey," *Artificial Intelligence Review*, vol. 55, no. 2, pp. 945–990, Feb. 2022, doi: 10.1007/s10462-021-09997-9

[12] V. Mnih et al., "Human-Level Control Through Deep Reinforcement Learning," *Nature*, vol. 518, no. 7540, pp. 529–533, Feb. 2015, doi: 10.1038/nature14236

[13] K. Arulkumaran, M. P. Deisenroth, M. Brundage, and A. A. Bharath, "Deep Reinforcement Learning: A Brief Survey," *Institute of Electrical and Electronics Engineers Inc.*, Nov. 01, 2017, doi: 10.1109/MSP.2017.2743240

[14] V. Shah, "Revista Española de Documentación Científica Reinforcement Learning for Autonomous Software Agents: Recent Advances and Applications," doi: 10.5281/zenodo.10779477

[15] R. Sharma, "Lyapunov Theory Based Stable Markov Game Fuzzy Control for Non-Linear Systems," *Engineering Applications of Artificial Intelligence*, vol. 55, pp. 119–127, Oct. 2016, doi: 10.1016/j.engappai.2016.06.008

[16] L. Buşoniu, R. Babuška, and B. De Schutter, "A comprehensive survey of multiagent reinforcement learning," *IEEE Transactions on Systems, Man, and Cybernetics, Part C (Applications and Reviews)*, *38*(2), 156–172, doi 10.1109/TSMCC.2007.913919

[17] M. Lauri, D. Hsu, and J. Pajarinen, "Partially Observable Markov Decision Processes in Robotics: A Survey," *IEEE Transactions on Robotics*, vol. 39, no. 1, pp. 21–40, Feb. 2023, doi: 10.1109/TRO.2022.3200138

[18] H. Kurniawati, "Partially Observable Markov Decision Processes and Robotics," *Annual Review of Control, Robotics, and Autonomous Systems 2022*, vol. 50, pp. 253–277, 2024, doi: 10.1146/annurev-control-042920
[19] M. T. J. Spaan, "Partially Observable Markov Decision Processes," 2012, pp. 387–414. doi: 10.1007/978-3-642-27645-3_12
[20] L. Villalobos-Arias, D. Martin, A. Krishnan, M. Gagné, C. M. Potts, and A. Jhala, "Modeling Risk in Reinforcement Learning: A Literature Mapping," Dec. 2023 [Online]. Available: http://arxiv.org/abs/2312.05231
[21] K. Arulkumaran, "*Sample Efficiency, Transfer Learning and Interpretability for Deep Reinforcement Learning.*" (Doctoral dissertation, Imperial College London).
[22] O. Singh, V. K. Barodiya, A. Paridwal, & A. Makhija (2020). Reinforcement Learning in Robotics: Challenges and Applications. *Turkish Journal of Computer and Mathematics Education (TURCOMAT)*, vol. 11, no. 2, pp. 743–745. 10.61841/turcomat.v11i2.14417

Chapter 6

High performance robotic arm controller using FPGA

B. Khaleelu Rehman, B. Iftekhar Hussain, P. Anuradha, Kalyan S Kasturi, and Mohammed Abdul Nasar

6.1 INTRODUCTION

FPGA based robotic arm controllers restore or enhance the functionality of the human body, providing greater independence and improving the quality of life for those in need. The FPGA's feature of reprogrammability ensures adaptability to different robotic arm configurations, while parallel processing efficiently executes complex control algorithms. These controls provide accurate results after integration of different hardware peripheral devices. This work endeavors to design an FPGA-based robotic arm controller with the key objective of achieving unparalleled precision and control over the movements of the robotic arm.

6.2 LITERATURE SURVEY

Urmila Meshram *et al.* discussed hardware and software co-design for a robot arm controller using an FPGA chip, focusing on position control. The system comprises a closed-loop controller, H-bridge driver circuit, DC motor, encoder, and a five-degrees-of-freedom robot arm model. VHDL coding is employed for designing modules like PWM generator, DC motor driver logic, memory controller, 8-bit register, and rotary encoder. PWM is utilized for digital speed control of the DC motor, implemented through a counter and comparator [1].

Meshram & Harkare, proposes 5 axis robotic arm controller using FPGA. They had outlined the design and implementation of a robotic arm controller utilizing an FPGA chip comprising four key components—controller system, arms, drive circuit, and sensors. The system employs FPGA-generated control signals for each motor, with position feedback from shaft encoders sent back to the FPGA leveraging the Spartan-II FPGA's flexibility, re-configurability, and capability to handle slow clocks for significant delays. The paper emphasizes motion control for stepper motors.

 DOI: 10.1201/9781003568995-6

Successful programming, simulation, and interfacing of the ULN2803A with the FPGA are achieved. The advantages of FPGA-based control systems over DSPs and microcontrollers include enhanced performance, flexibility, cost-effectiveness, and faster time-to-market. The stepper motor driver circuit is pivotal, enabling controlled current and flux direction changes. Precise position control is achieved through the implementation of a shaft encoder, translating motor rotations into pulses for closed-loop control of speed and angular displacement [2].

Kung and Shu had shown the results through experiments for development of a FPGA-based motion control IC for robot arm [3]. FSM Alkhafaji *et al.* had focused on the application of using FPGA as a development reconfigurable platform for a controller solution [4]. Kung *et al.* had portrayed the implementation of inverse kinematics and servo controller for robot manipulator using FPGA [5]. Wan *et al.* had analyzed the software and hardware optimization techniques and main technical issues in FPGA-Based Robotic Computing [6]. Martinez-Prado *et al.* designed a field-programmable gate array (FPGA)-based motion controller for robotic applications [7]. This chapter designs an FPGA-based robotic arm controller with the key objective of achieving unparalleled precision and control over the movements of the robotic arm.

6.3 BOOLEAN FPGA BOARD

The board is centered on a Xilinx Spartan-7 FPGA (XC7S50) shown in Figure 6.1 that can be configured into a virtually unlimited number of hardware circuits [8]. It includes a large collection of peripheral devices and ports, allowing countless designs to be built without the need for any other components. The Spartan-7 FPGA includes:

- More than 8,000 FPGA slices, each containing four 6-input look-up tables and 8 flip-flops, equivalent to 52K logic cells;
- 120 DSP blocks, each with dual 24-bit adders, a 2's compliment multiplier, and a 48-bit accumulator;
- 337Kbytes of high-speed, dual port RAM;
- Five clock management tiles that can generate a wide variety of clock signals from a single outside source;
- A 12-bit, 1MSPS analog-to-digital converter and many other circuits and functions.

It features memory with a 128 Mbit Serial Flash shown in Figure 6.2, that displays including two 4-digit 7-Segment displays and HDMI source, audio channels, switches, LEDs, push-buttons, clocks, expansion ports, and miscellaneous hardware components, making it a versatile and powerful platform for digital circuit and system development.

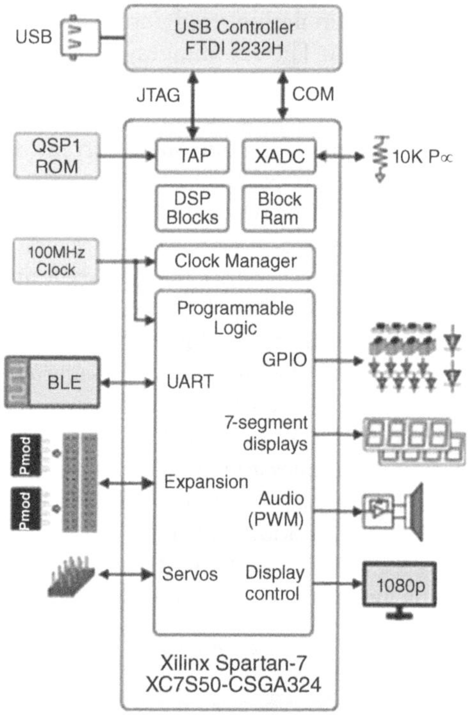

Figure 6.1 Architecture of Boolean board [9].

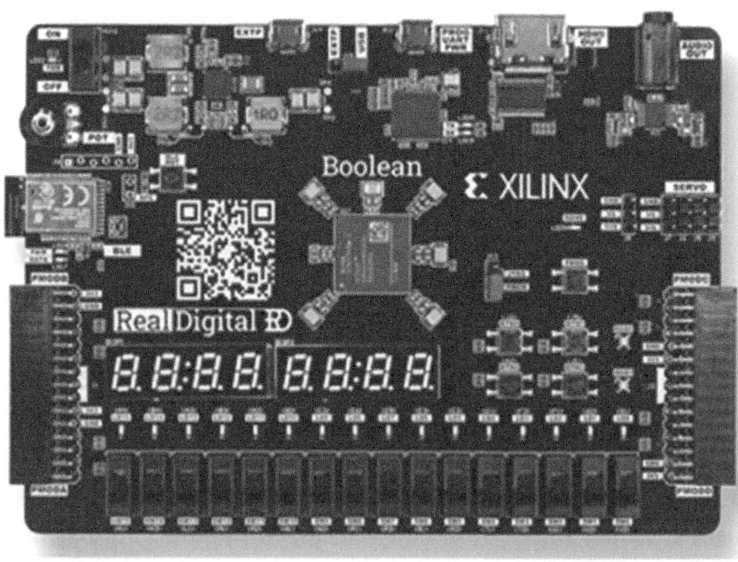

Figure 6.2 Boolean board [9].

6.4 SERVO MOTORS

Typically, a 50 Hz PWM signal with a pulse width ranging between one m to two m is employed to specify the desired position or angle of the servo motor's shaft, providing precise and customization control capabilities.

6.4.1 Why servo motor?

When it comes to robotics and automation, servo motors shown in Figure 6.3 are often mentioned. Whatever we want to do with robotics, we'll always be able to locate a servo motor that matches the bill because: Servo motors are available in a variety of shapes and sizes [9]. Servo motors differ in terms of power and torque. For use in larger applications such as assembly lines, torque is higher in larger servos. Torque is typically reduced in lightweight applications such as packing. Companies must use economical and cost-effective choices to keep the cost of these robotic arms down, and servo motors fit the bill well. All servo motors require a voltage and PWM (pulse-width-modulation). Servo motors shown in the figure are much lighter than stepper motors since there is no motor control to worry about, and we only need to connect three wires instead of the six that stepper motors require [10]. Tzou and Kuo had designed and implementation of a motor control IC for permanent magnet AC servo motors using the FPGA [11].

The lighter weight allows robotics and robot arms to move around more freely while maintaining equilibrium. Servo motors move very precisely, making them an excellent choice for applications with tight tolerances. Servo motors are widely available, with most industrial parts dealers carrying them. This is a significant benefit for companies looking for a solution that will not annoy them when searching for replacement.

In the realm of robotic arm control, leveraging FPGA technology as the central component offers a versatile and dynamic approach. The FPGA chip serves as the brain, providing real-time control and adaptability. Through

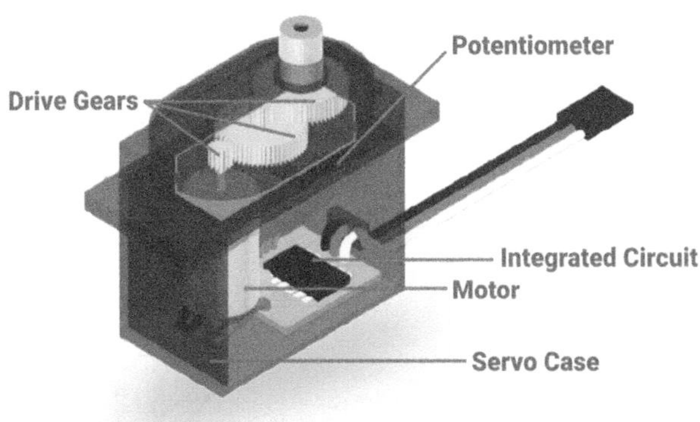

Figure 6.3 Servo motors [11].

Verilog coding, essential hardware components and PWM generators are seamlessly integrated [12]. The FPGA's reconfigurable nature allows for precise control of servo motors, influencing the arm's movement. Rotary encoders offer crucial position feedback, enabling accurate and responsive adjustments. The FPGA's ability to merge hardware and software functionalities enhances the efficiency of control systems. This hardware-software co-design facilitates bidirectional motor control and ensures optimal utilization of FPGA capabilities [13]. The result is a streamlined and flexible robotic arm controller, capable of intricate movements and precise positioning.

6.5 FPGA INTEGRATION WITH ROBOTIC ARM

The second servo motor is positioned at the joint connecting the forearm to the base, controlling the elbow joint movement. The third servo motor is mounted at the end of the forearm to control the gripper or wrist movement, completing the 3 DOF configurations.

With the servo motors mounted, the segments of the robotic arm can be connected together using screws and nuts. This step involves assembling the arm structure while ensuring that the joints allow the desired range of motion without any mechanical binding. Special attention should be given to the connection between the rotating base and the forearm, ensuring it allows smooth rotation without constraints.

After the mechanical assembly is complete, as shown in Figure 6.4 the next phase involves wiring the servo motors to the controller for control. Use male-to-female jumper wires to connect each servo motor, identifying

Figure 6.4 Robotic arm.

and connecting the signal (typically orange or yellow), power (red), and ground (brown) wires to the corresponding pins on the Arduino [14]. Provide power to the servos either through the Arduino's 5V output or a separate power supply to the servos' power and ground lines.

Lastly, prioritize safety considerations throughout the construction process and operation of the robotic arm. Always handle the arm with caution, especially during testing and task execution. Implement safety features such as emergency stop buttons or software limits to prevent unintended movements or damage.

6.6 IMPLEMENTATION OF THE SYSTEM

Write a Verilog Hardware description language code to implement three modules i.e.,

- Angle to on-time converter,
- PIPO shift register, and
- PWM generator.

Collectively, these modules form a system designed for generating PWM signals based on angle and trigger inputs. Let's discuss each module with detailed pin description and simulation waveforms on X-Sim simulator

Angle to on-time Module:

- If the input angle doesn't match any specified cases, the default on-time is set to 0.
- This module is likely integral to a larger system, contributing to the generation of PWM signals or the timing control of specific operations based on selected angles. Figure 6.5 shows the Duty cycle of PWM signals.

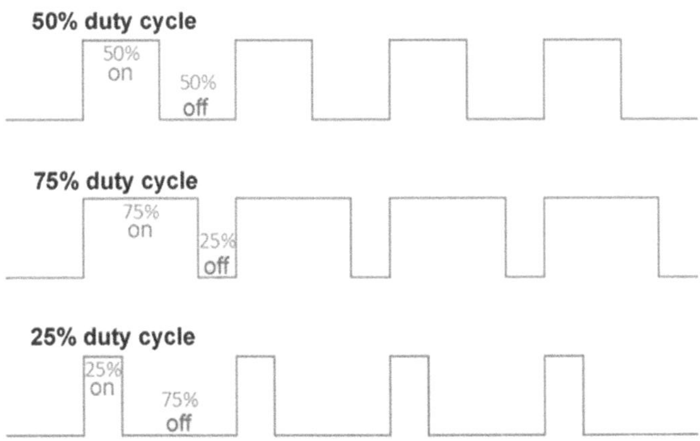

Figure 6.5 Duty cycle of PWM signals.

6.7 RESULTS AND DISCUSSION

The Boolean board, featuring a Spartan-7 FPGA, serves as the control center for the robotic arm. It responds to given inputs by directing the arm's movements accordingly. Each input angle is translated into corresponding on-time values, which are then converted into PWM (Pulse Width Modulation) signals sent to the arm's motors. When the PWM signal shown in Figure 6.5 is high when the motors rotate until reaching the desired position within that time frame.

6.8 RTL SCHEMATIC

The RTL schematic shown in Figure 6.6 provides a clear representation of the digital components within the design. This view makes it easy to comprehend the implemented logic and understand how different elements interact.

Below is the RTL schematic view depicting the project's implementation.

6.9 SIMULATION RESULTS

The PWM signal output shown in I-Sim waveform Figure 6.7 is given to the motors in the robotic arms [15]. PWM signals are essential for controlling motors in robotic arms. By adjusting the duty cycle of the PWM signal, the motor's speed is regulated — a higher duty cycle increases speed.

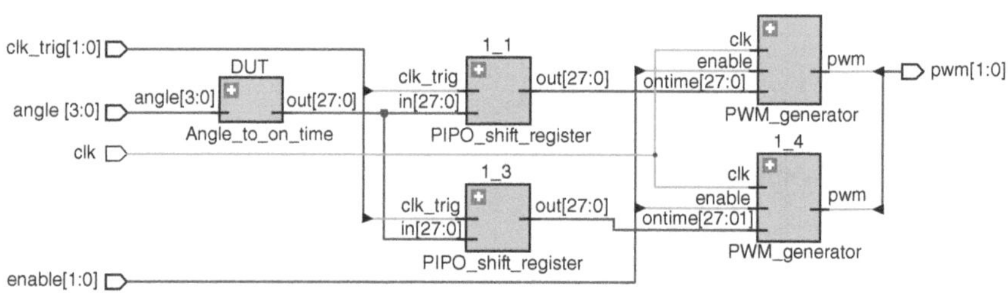

Figure 6.6 RTL schematic view.

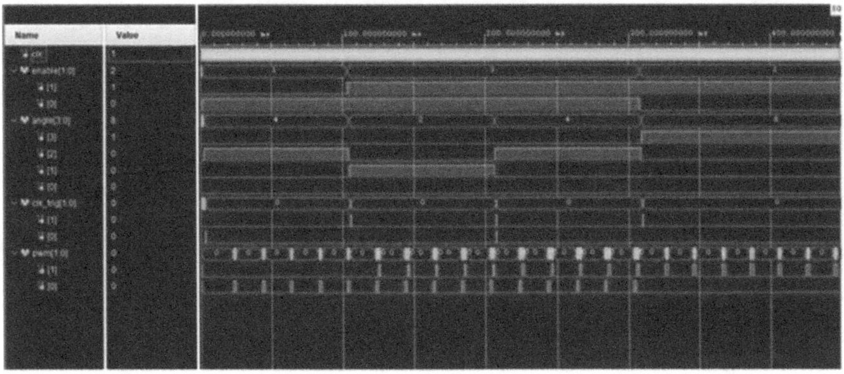

Figure 6.7 Expected PWM Signals to control the motors in robotic arm.

6.10 RESOURCE UTILIZATION

Here are the total resources of Boolean Board which are used to implement the robotic arm.

- The total available Slice LUTs in Boolean board are 32600 of which we used only 53 LUTs and each module has utilized different number of LUTs
- The total available Slice registers in Boolean board are 65200 of which we used only 78 LUTs and each module has utilized different number of LUTs [16]
- The total available Bonded IOBs and BUFGCTRL in Boolean board are 210 and 32 of which we used 11 Bonded IOBs and 32 BUFGCTRLs.

Figure 6.8 is the synthesis report generated using Xilinx Vivado. The LUTs, which are programmed to execute combinational logic functions, are currently using 53 out of 32,600 available, indicating a utilization rate of 0.16%. The FFs, synchronous storage elements, are at 0.12% utilization, with 78 out of 65,200 being used. Registers, which can include FFs, specifically refer to 7,810 FFs in this context.

Regarding slice logic and memory, the LUTs are allocated for logic operations (53 out of 32,600 slices) but not for memory functions (0 out of 9,600). All 78 registers are being utilized as flip-flops for storage purposes. Additionally, the design incorporates F7 Muxes (1 out of 16,300) and F8 Muxes (1 out of 8,150) to manage data routing within the FPGA, shown in Figure 6.9 contributing to the system's overall functionality.

Name	Slice LUTs (32600)	Slice Registers (65200)	Bonded IOB (210)	BUFGCTRL (32)
Top_Module	53	78	11	3
DUT (Angle_to_on_time)	5	0	0	0
I_1 (PIPO_shift_register)	3	10	0	0
I_2 (PWM_generator)	21	29	0	0
I_3 (PIPO_shift_register_0)	3	10	0	0
I_4 (PWM_generator_1)	21	29	0	0

Resource	Utilization	Available	Utilization %
LUT	53	32600	0.16
FF	78	65200	0.12
IO	11	210	5.24

Figure 6.8 Synthesis report.

Site Type	Used	Fixed	Prohibited	Available	Util%
Slice LUTs*	53	0	0	32600	0.16
LUT as Logic	53	0	0	32600	0.16
LUT as Memory	0	0	0	9600	0.00
Slice Registers	78	0	0	65200	0.12
Register as Flip Flop	78	0	0	65200	0.12
Register as Latch	0	0	0	65200	0.00
F7 Muxes	0	0	0	16300	0.00
F8 Muxes	0	0	0	8150	0.00

Figure 6.9 Resource utilization report.

6.11 POWER ANALYSIS OF THE DESIGN

Figure 6.10 shows the power analysis using the X-power analyzer software tool [16]. Dynamic Power: 0.421 watts (85%) Signals: 0.214 watts (51% of total power, 51% of dynamic power) Logic: 0.171 watts (41% of total power, 41% of dynamic power) Static Power: 0.073 watts (15% of total power)

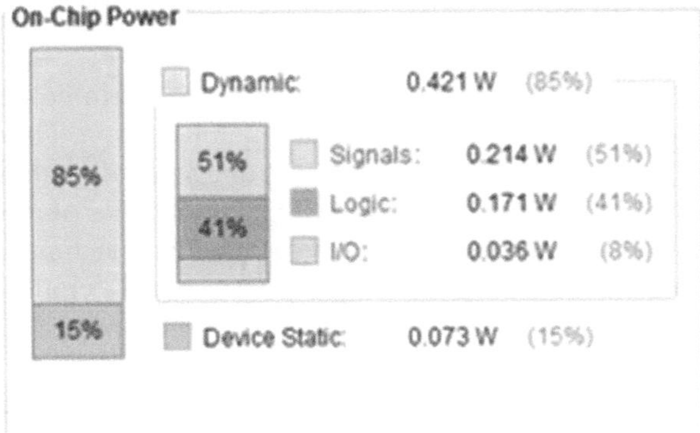

Figure 6.10 Power utilization report.

6.12 FINAL DESIGN OF THE FPGA INTEGRATION OF THE ROBOTIC ARM

The final output is shown in Figure 6.11 is the connections with motor and FPGA. Figure 6.12 describes after integrating the robotic arm with the Spartan 7 Boolean board of FPGA and the servo motor.

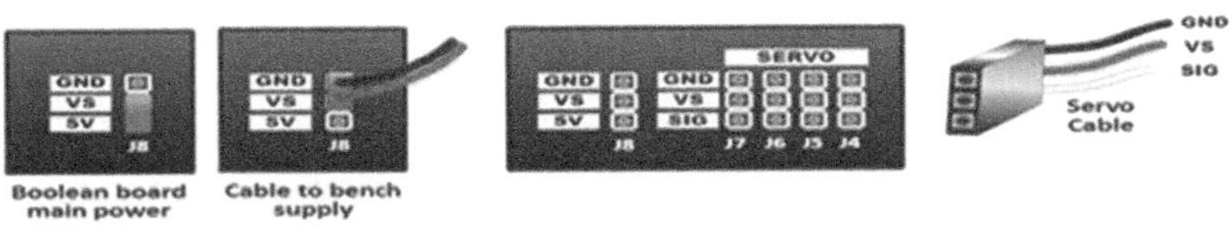

Figure 6.11 Boolean board connection to the robotic arm.

Figure 6.12 Final prototype.

6.13 CONCLUSION

In conclusion, the integration of FPGAs in robotic arm controllers represents a transformative step in the evolution of automation. The combination of real-time processing, parallel computation, and adaptability positions FPGAs as a robust platform for handling the complexities of modern robotic systems. The future developments outlined, including machine-learning integration, advanced vision capabilities, haptic feedback, cloud connectivity, and collaborative swarm robotics, collectively point towards a new era of intelligent and versatile robotic arms. As technology continues to progress, the synergy between FPGA-based controllers and emerging technologies is likely to redefine the landscape of automation, driving innovation across industries and opening up new possibilities for robotic applications in diverse fields.

ACKNOWLEDGMENT

The authors would like to sincerely thank the Ministry of Electronics and Information Technology (MeitY) for providing necessary hardware/software support for this work under the Chip to Start-up Program being implemented at CBIT, Hyderabad. Further the authors thank the students Ms. M. Vidya. Mr.J.Nishanth and Ms. M.Akshita for their continued support in implementing the design.

REFERENCES

[1] Urmila Meshram, Pankaj Bande, and R.R. Harkare. "Hardware and Software Co-design for Robot Arm Position Control Using VHDL & FPGA" *International Conference on Advances in Recent Technologies in Communication and Computing*, 2009, pp. 780–782, IEEE Xplore.

[2] Urmila Devendra Meshram and R. R. Harkare. "FPGA based five axis robot arm controller." *International Journal of Electronics Engineering* 2(1) 2010: 209–211.

[3] Ying-Shieh Kung, and Gua-Shieh Shu. "Development of a FPGA-based motion control IC for robot arm." *IEEE International Conference on Industrial Technology*. IEEE, 2005.

[4] Falih Salih Alkhafaji, Wan Z. W. Hasan, Mazlina. M. Isa, and N. Sulaiman, "Robotic controller: ASIC versus FPGA—A review." *Journal of Computational and Theoretical Nanoscience* 15, 1–25, (2018).

[5] Ying-Shieh Kung, Kuan-Hsuan Tseng, Chia-Sheng Chen, Hau-Zen Sze, and An-Peng Wang, "FPGA-implementation of inverse kinematics and servo controller for robot manipulator" *2006 IEEE International Conference on Robotics and Biomimetics*. IEEE, 2006.

[6] Zishen Wan, Bo Yu, Thomas Yuang Li, Jie Tang, Yuhao Zhu, and Yu Wang, "A survey of FPGA-based robotic computing." *IEEE Circuits and Systems Magazine*, 21, 2 (2021): 48–74.

[7] Miguel Martinez-Prado, Alfonso Franco-Gasca, Gilberto Herrera-Ruiz, and Otoniel Soto-Dorantes, "Multi-axis motion controller for robotic applications implemented on an FPGA." *The International Journal of Advanced Manufacturing Technology* 67 (2013): 2367–2376.

[8] Vivek Ramakrishnan, Nalamwar Sanchit Gopal, Rahul Ashok, and S. Moorthi "FPGA based DC servo motor control for remote replication of movements of a surgical arm", *TENCON 2011, IEEE Region 10 Conference*. IEEE, 2011.

[9] https://www.amd.com/en/corporate/university-program/aup-boards/realdigital-boolean-board.html

[10] B. Khaleelu Rehman, Ramalla Isaac, K. Abdul Munaf, Salauddin Mohammad, and Mudassar Basha, "FPGA implementation of area-efficient binary counter using Xilinx IP Cores." *Sustainable Infrastructure Development: Select Proceedings of ICSIDIA 2020*. Part of the book series: Lecture Notes in Civil Engineering (LNCE, volume 199). Springer, Singapore, 2022, pp. 147–156.

[11] Ying-Yu Tzou, and Tien-Sung Kuo. "Design and implementation of all FPGA-based motor control IC for permanent magnet AC servo motors." *Proceedings of the IECON'97 23rd International Conference on Industrial Electronics, Control, and Instrumentation* (Cat. No. 97CH36066). Vol. 2. IEEE, 1997.

[12] B. Khaleelu Rehman, G. Vallathan, Vetriveeran Rajamani, Mudasar Basha, and Raman Kumar, "Area efficient high-speed binary divider using Xilinx IP core." *AIP Conference Proceedings*. Vol. 2971. No. 040042. AIP Publishing, 2024.

[13] Ahmed R. Ajel, Huda M. Abdul Abbas, and Mohannad Jabbar Mnati. "Position and speed optimization of servo motor control through FPGA." *International Journal of Electrical & Computer Engineering (2088–8708)* 11(1) (2021): pp. 319–327.

[14] B. Khaleelu Rehman, G. Vallathan, Vetriveeran Rajamani, and Salauddin Mohammad, "Xilinx FPGA and Xilinx IP cores: A boon to curb digital crime." *Digital Forensics and Internet of Things: Impact and Challenges* (2022): 131–162.

[15] Mohamad Khairi Ishak, Ho Kar Hwai, and Lee Kong Lam. "Real time FPGA-based ethernet control communication for robotic arm." *MATEC Web of Conferences*. 95, EDP Sciences, 2017.

[16] Jung Uk Cho, Quy Ngoc Le, and Jae Wook Jeon. "An FPGA-based multiple-axis motion control chip." *IEEE Transactions on Industrial Electronics* 56.3 (2008): 856–870.

Chapter 7

New-age innovative fusion of drones and smart technology

Akanksha Singh, Lipsa Das, and Manjeet Singh

7.1 INTRODUCTION

Unmanned Aerial Vehicles, popularly referred to as Unmanned Aircraft Systems, are those systems developed during the 20th century, but recently because of the extreme flexibility in their applications they have lately acquired popularity [1, 2]. These systems are preferred because of their high efficiency in executing real-time operations, flexibility, and cheaper acquisition costs. Since its invention, UAV technology has been developed remarkably. Some of the examples are that there are reductions in weight, increased lifetime for the batteries, use of smart sensors and 4K cameras onboard, and use of machine learning and deep learning for increased autonomy during operations. The Federal Aviation Administration's definition of "Unmanned Aircraft System" is widely known, and the other terms it uses interchangeably are "Flying Robot," "Pilotless Aircraft," and "Remote Plane." But the most widely used term is taken from the military nomenclature and is "Drone." The FAA Modernization and Reform Act of 2012 defines an "Unmanned Aircraft" as any aircraft operated without a human being inside or in direct control of the aircraft whether onboard or remotely [3, 4]. A UAV can be considered as a "smart aircraft system" which remotely controls or operates through its onboard computer system. Its major functions comprise mainly real-time monitoring video capture and other operational activities. UAVs are part components of UAS, whereby essential elements comprise controllers, communication systems, sensors, cameras, and other mechanical parts. Figure 7.1 shows the growth in UAV market size from 2017 to 2027. The growth in global UAV adoption is entering other markets; the most significant being the adoption of agricultural UAVs and media, oil, gas, e-commerce, among others. This review will encompass all aspects regarding the transformation in drone technology, their application, and the impact of Artificial Intelligence. Development in AI greatly enhances drones through navigation, analysis of data, and fields like agriculture, environmental monitoring, and healthcare. This nexus provides an avenue for this review to

 DOI: 10.1201/9781003568995-7

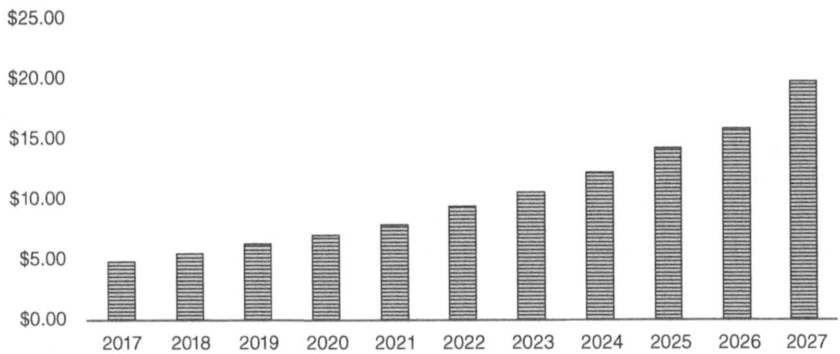

Figure 7.1 UAV drone market size ($BN USD).

clarify the present stage of AI-driven drone technology, trends in emergence, and possibly future developments [5]. It is known that once the integration of AI and drone technology is understood, professionals, researchers, policymakers, or enthusiastic minds would love to know about the possibilities and challenges that come with merging those technologies together in ways that have enormous implications for revolutionizing industries at large [6]. It basically marks a turning point in the development of unmanned aerial systems with AI, giving scope to unprecedented applications with opportunities arising from the very foundations for more developments [7].

7.2 HISTORICAL DEVELOPMENT

Deeply engrafted within the development of UAV technology are the historical steps regarding artificial intelligence (AI), and it's merging into drone technology. UAV origins may trace back to the start of the 20th century when early experiments in the development of remotely controlled aircraft were started [8, 9]. This remained rather a peripheral affair of this sector until the mid-20th century, and the significant momentum given by the Cold War during the 20th century. Reconnaissance drones such as the Ryan Model 147 appeared in the 1960s. They were used mainly for surveillance purposes by the United States. These primitive drones were quite basic and utilized minimal autonomy with very simple radio-control systems [10]. Technology of the late 20th century brought a new era in UAVs. The advancement in the field was noted in better navigation. These innovations enabled the usage of AI in drones, meaning the function of drones had changed. Drones were now applied to meet military goals by collecting reconnaissance and intelligence. They came without the cost of human lives.

Early generations were, however, encumbered with manual-control dependency and a somewhat limited autonomy. Meanwhile, technological barriers, regulatory obstacles and a lack of social acceptance have limited their use almost exclusively to the military [10, 11]. In the civil industry, they were primarily used for agriculture in the form of crop monitoring as well as search and rescue missions to gather environmental data. Yet they are not optimized to the extent that they can perform wondrous things without AI-integration that made a paradigm shift possible. By the end of the twentieth century and into the twenty-first, drones had incorporated simple AI features, granting them the ability to fly autonomously along programmed flight paths and perform basic actions. Early AI integrations mainly improved navigation and control, representing the first steps toward autonomous drone operations [13]. The mid-2000s marked a significant turning point as AI advancements allowed drones to achieve intelligent autonomy. They could process data in real time and could take decisions independently, change their flight paths from predefined flight paths to dynamic adaptive operation. Using the principles of machine learning, they became more cognitive with a recognition pattern, analysis of sensor data, and able to respond in real-time to environmental changes. Their functionality became more than surveillance-based with this advancement; they were utilized in complex missions. Computer vision even changed the face of drones during the 2000s and 2010s. With new technology consisting of cameras, sensors, and algorithms, drones started looking at their surroundings with narrow interpretations of spaces and objects capes to phenomenal detail of 3D space to view objects' locations on the ground concerning the drones' own locations. Drones could now recognize things, develop maps of irregular surfaces, and even detect collision zones. All this enhanced operational safety and efficiency among many applications in search and rescue, infrastructure inspections, and environmental monitoring [12]. In the mid-2010s, the application of machine learning into neural networks revolutionized the already available techniques they had. Algorithms are sophisticated enough to allow crane teaching by data and adapt themselves to newer environment changes improved the performance over time. Machine learning allowed decision-making to be data-based; deep-learning models provide further techniques for image recognition through which they could recognize objects. These advancements led to an increase in channels for precision agriculture, such as using drones that can read the status of crops and find more efficient ways of farming [14, 15]. Drones have come a long way, from monarchically controlled kite-like flying objects to highly autonomous entities powered by artificially intelligent algorithms. This has opened new territories in military and civilian applications. Such synergy offers wholly new dimensions to production and further innovations in the future of UAVs across a variety of industries [16]. Table 7.1 presents the applications of drone technology across various disciplines.

Table 7.1 The convergence of drones and smart technology across various applications, features, and benefits

Application area	*Smart technology features*	*Drone capabilities*	*Benefits/Impact*
Agriculture	IoT-enabled sensors for soil and crop monitoring	Aerial imaging and precision spraying	Increased crop yield, reduced resource wastage
Infrastructure Inspection	AI-based fault detection algorithms	High-resolution cameras and thermal imaging	Cost-effective and safer inspections
Disaster Management	Real-time analytics and communication networks	Rapid deployment and payload delivery	Faster response times, saved lives
Urban Planning	3D modeling and data visualization tools	Mapping and data collection capabilities	Smarter urban designs, efficient resource use
Logistics and Delivery	Route optimization and AI-driven automation	Autonomous flight and precise navigation	Faster deliveries, lower emissions
Environmental Monitoring	Remote sensing and data analysis tools	Long-range endurance and pollutant detection	Improved conservation efforts, reduced pollution

7.3 LITERATURE REVIEW

7.3.1 Educational advantages of drone technology

These UAVs are equipped with state-of-the-art toolkits, including robotic arms and high-definition cameras; for example, Amazon introduces the use of drones as product delivery devices in a corporation setting, agricultural farmers employ drones in such activities as monitoring animals, irrigation system management, and pesticide application, firefighters do also use drones to conduct surveys of fire-damage conditions to enhance situation awareness [17]. The increased use of drones has also been embraced into the education sector where these drones are gaining high influence as novel teaching resources [18, 19]. According to Walsh (2020), drones can prove to be effective for learning activities in seven activities of logic and deductive reasoning, debate, geography, higher mathematics, electronics, and hand-eye coordination [20]. Cenejac (2021) added that images taken from drones could trigger thinking inside the minds of students in speaking and writing classes and enhance exercise schemes in physical education [21]. With their potential to support and enrich student learning, drones are increasingly considered an alternative instructional tool that can change the face of

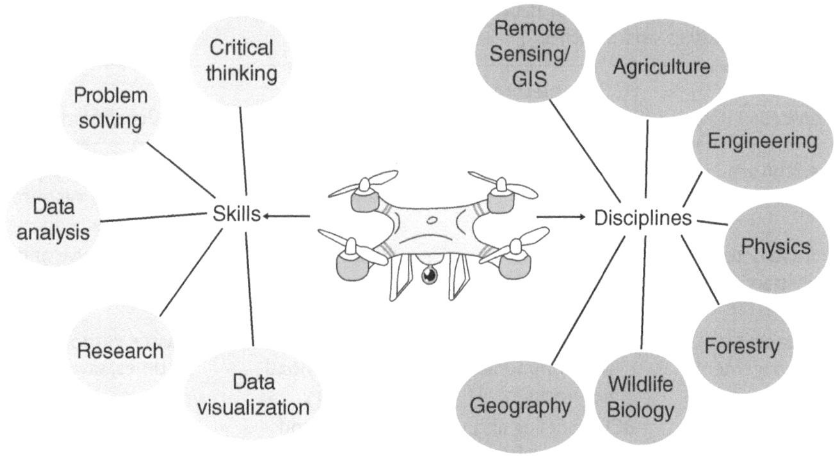

Figure 7.2 Educational advantages of drone technology.

traditional learning environments. Yet, such promising applications would require much more research on whether including drones in a classroom will be better in terms of learning outcomes compared to traditional teaching methods [22]. Figure 7.2 shows the educational advantages of drone technology.

7.3.2 Limited research in drone programming

Drones for learning can be categorized into heavyweight drones (over 1 pound) and lightweight drones (under 1 pound). The heavyweight drones like DJI Mavic are better suited for outdoor learning as they can withstand the wind better [23]. Lightweight drones like the Parrot Mambo are better suited for indoor learning as they are easier to handle in small spaces [24]. Both types of drones come with remote controls, but lightweight drones often come with programming capabilities making them great tools for teaching coding and computational thinking. According to Hussey and Wakefield, several elementary schools in the US have introduced lightweight drones into the classroom and students are programming drones for indoor flights [25, 26]. They have seen motivation among students but few educational assessments on drone programming. Most research on drones in education is theoretical and limited empirical studies have provided little data in the form of quantitative or qualitative. For example, Schaeffer and Olson developed a graduate information technology course that explores theoretical applications and development of drone technology [27]. Students liked the course, but no formal assessments were made. Similarly, Carnahan and Crowley proposed that curricula integration of drones would encourage student engagement but did not provide data [28]. In an educational experiment by Palaigeorgiou, Malandrakis, and Tsolopani, one group of students went on a field trip while another group watched drone captured videos of the same location from a lab. Students who witnessed the drone footage enjoyed it and liked the view the drone gave them of the whole field [29]. But despite these results which seem promising for drones in education, more research is needed to evaluate their impact on learning outcomes.

7.3.3 Enhancing spatial visualization abilities

Spatial visualization is the ability to manipulate objects in your mind as if they were 2D or 3D objects. Scientific training improves this skill [30]. Traditionally the tools used to teach spatial visualization were plastic models [31]. Now the tools are 3D multimedia software. Research has shown that dynamic geometry and open source 3D software can help students develop spatial understanding [32]. Drones are an alternative to traditional and virtual methods because they offer hands on 3D learning. While students watch virtual objects on a screen, operating a drone lets them physically interact with their surroundings [33]. With remote controls or programming they can navigate drones and design flight paths, exercise spatial reasoning and interpret 3D geographical information [34]. Integrating drones into the curriculum has huge potential for developing spatial visualization skills by combining practical engagement with technology [35].

7.3.4 Enhancing sequencing abilities

Sequencing is the procedure of putting objects or planning actions in a particular order [36]. This kind of skill is often seen in a child's activity like sorting or retelling stories in a particular sequence. Sequence is also somewhat of a link to the development of math and problem-solving skills [37]. According to Sarama and Clements, the usage of building blocks by children is a clear manifestation of the basic mathematical thinking [37]. In the same vein, the studies by Zelazo and company deal with sequencing as a basic part of the planning and, therefore, the problem-solving framework. Computer programming is a great way to improve proficiency in sequencing [38, 39]. It is when the student codes commands in a line that is arranged in the order of the outcome desired [40]. Visual block programming languages like Scratch, which are typically seen in elementary schools, help kids through matching up color-coded blocks to create animations [41]. Brennan and Resnick described this deductive process as a crucial aspect of computational thinking [42]. A study by Kazakoff and colleagues showed that kids who underwent programming activities and worked on putting blocks in various patterns to solve tasks observed a noticeable improvement in their sequencing skills [36]. With this learning method, we can see how computer programming manifests a situation where students are oriented with sequencing skills and have hands-on experience, eventually learning logical organization and problem-solving.

7.3.5 Integrating smart technology into a sustainable curriculum

UNESCO's "Teaching and Learning for a Sustainable Future" program place the integration of key instructional themes into the school curriculum as the central emphasis in this program. Among these, "future perspectives" and "citizenship education" are part of it that are given much importance. Incorporating smart technologies under the policy of developing a sustainable curriculum is a systemic transformation, which results in the establishment of a digital learning environment, in which students are being empowered through the recent trend of digital citizenship [43]. The educators are promoting communication skills and steering students through the application of fledgling digital tools to investigate and surmount learning difficulties. The use of drones, as a form of "smart digital technology," not only provides a new dimension to the development of creativity and critical thinking in the classrooms. They let our children be drone pilots by allowing them to learn some basic programming and through interactive processing activities to become computational learners. These experiences make the students ready for an interesting life in a highly technologically developed society [44]. Table 7.2 highlights the growth and impact of integration of drones and smart technology.

Table 7.2 Growth and impact of drones and smart technology integration

Metric	*2018*	*2020*	*2025 (Projected)*	*Growth (%)*
Global Drone Market Size (USD Billion)	14.1	22.5	63.6	351
Adoption Rate of Smart IoT Drones (%)	15	35	70	367
Average Cost Reduction in Agriculture (%)	10	20	40	300
Infrastructure Inspection Efficiency (%)	50	70	90	80
Delivery Time Reduction (in Hours)	4	2.5	1	–75
Annual Carbon Emission Savings (MT)	0.5	1.2	5	900

7.4 THE ROLE OF MACHINE LEARNING IN SMART DRONES

Predictive maintenance is one of the applications in drone technology that prevents failures before they become more expensive. The AI algorithms have identified patterns and anomalies that indicate possible problems or issues, including mechanical failure, battery degradation, or even software flaws. Telemetry data, sensor reading, and flight logs let the machine-learning models point out early symptoms of trouble for operators to take preventive measures. Thus, this approach prevents even more downtime, the drones last longer, and the cost of operation is lower [45]. The integration of machine learning for predictive maintenance enhances drones' reliability and efficiency. Drones will become very consistent in their performance if critical issues are addressed before becoming major problems, thereby averting the chances of probable unplanned disruptions. AI models can optimize schedules for performing maintenance, giving priority to critical repairs and even anticipate the lifespan of different parts, thus always keeping drones in the best working condition [46]. That reduces unexpected breakdowns, maximizing operational uptime. Machine learning also serves to make sense of all the drone-generated data it collects after flying. Through AI-driven algorithms, sensor data, image data, and telemetry, streams can be interpreted in near real-time. This then gives insights into, for instance, land surveying or mapping and monitoring environmental studies [47]. These algorithms allow a drone to immediately and accurately identify objects, classify characteristics of its terrain, or detect several environmental changes much more effectively than has been possible using other techniques [48]. Environmental monitoring can be done with the help of AI-based drones to observe biodiversity, ecosystems monitoring, and habitat changes assessment. Machine learning models in the analysis of aerial images assess the trends of vegetation, populations of wildlife, and changes to the environment due to human activities. This technology is applied in conservation activities, land management, and research in science [50]. In short, the predictive maintenance and data analysis of drones through machine learning are developed as described. It would allow AI drones to get early detection of system failure and handle a lot of data which would make it efficient and reliable. With advancement in technology, machine learning will open new avenues for the use of drones in several sectors, and this will change the face of applications like surveying, mapping, and environmental monitoring [49].

7.5 THE ROLE OF AI IN DRONE TECHNOLOGY

Autonomous navigation is the most prominent use of AI in drone technology, enabling drones to plan a route and alter it in real time. Drones would normally have routes programmed for them, but they lacked the ability to adjust the routes as they would encounter changing conditions. Data from onboard sensors like GPS, motion detectors and obstacle detection tools help AI-driven drones fly on their own. These systems look at weather, airspace rules and land features to plan the safest and most efficient routes [51]. AI flight planning allows drones to perform very well in complicated areas, making them useful for aerial checking, mapping and research.Obstacle detection and avoidance is another powerful AI feature. Manned aircraft aren't designed for complex or unpredictable environments [52]. But AI drones have advanced sensors and computer vision systems. They can pick up buildings, trees, and even other drones and fly around them. AI algorithms automatically process sensor data to identify new flight paths and avoid collisions when targeting targets. This is especially useful for applications such as search and rescue, where drones explore urban environments or tall vegetation. Find a problem person. AI helps drones recognize and follow objects [53]. A computer vision system with AI inside a drone reads visual data from cameras to spot and track things like cars, people or animals. Once found, drones follow the movement of these objects, giving good situational awareness for various uses like police work, border control and watching wildlife. With these skills, AI-powered drones help authorities more effectively follow suspects, secure borders or watch animals. In exploring resources and monitoring the environment, drones with AI are changing how data is gathered. Drones work with computer designs to find problems or unusual things in structures like bridges, pipelines and power lines [54]. These drones utilize AI algorithms to scan visual data and identify signs of wear or damage. These capabilities allow for pre-emptive maintenance and repair and save both time and resources. Drones can monitor ecosystem health, wildlife numbers, or land-use changes. The AI algorithms detect environmental hazards, monitor changes in habitats, and even detect illegal activities including poaching or deforestation [55].AI algorithms transform precision agriculture. With multispectral or hyperspectral cameras on drones, one can take high-resolution images of crops to monitor their health and growth patterns as well as yield. The AI algorithms process this data to create crop maps, pinpoint areas requiring attention and suggest interventions such as irrigation or fertilization. That way, the farmer can make a more informed decision to maximize yields, minimize input costs and reduce their environmental impact [56]. Precision spraying drones can easily be deployed to apply fertilizers, pesticides, or herbicides only on targeted areas, reducing waste and moving towards sustainable practices. So, in a nutshell, the applications of AI on drones begin from fully autonomous navigation and obstacle detection up to computer vision and precision agriculture. It unlocks various possibilities of autonomous flight with real-time obstacle avoidance and object tracking, as well as data-driven farming practices, to several industries. The advancement of AI will open more prospects of using the drones and hence increase new possibilities of innovation and efficiency [57].

7.6 CHALLENGES IN INTEGRATING SMART TECHNOLOGY WITH DRONES

The inclusion of intelligent technology in unmanned aerial vehicles presents a range of challenges. The integration of the drones' hardware and software is a significant technical hurdle that requires attention. From sensors to GPS systems, communication equipment, and flight controllers. It enables and ensures that these factors are precisely included within cutting-edge technologies like AI or machine learning. Most of the time, though, it's not possible [58]. The costliness of it is that the system only functions when something goes wrong in the system itself. Power consumption and battery life are also significant. Several of these features, such as AI-led obstacle avoidance, real-time data processing and autonomous navigation saves battery power on the drone's part. Why? This is a matter that requires investigation, particularly in relation to advanced drones that can enhance their capabilities for longer-range missions or more complex tasks. Safety and regulatory concerns are also important factors in the integration process. The safety of smart-tech drones must comply with aviation regulations and standards [59]. There are multiple questions about liability related to an incident involving AI and autonomous systems. As drones gain more intelligence, the management of airspace becomes increasingly intricate and necessitates new regulatory measures, for safe integration into the current air traffic system. Ultimately, security and data-related challenges occur. The most common cause of cyberattacks on drones is the sheer amount of data that advanced sensors and AI-based systems gather. Therefore, strong encryption is necessary to prevent sensitive information from being accessed [60].

7.7 FUTURE TRENDS IN DRONE AND SMART TECHNOLOGY INTEGRATION

Future drones will benefit from the added capability of integrated edge computing - meaning that the drone will process information on-site without the use of a cloud-based system [61]. This makes it more enhanced towards the decision-making process with higher speed and reliability for mission-critical applications [62]. Great collaborations by the academia, industry, and the government will be there to propel these developments ahead; however, enormous hurdles await them [63]. The success that is being witnessed also does not make such a move of integrating these technologies straightforward. It calls for considerable co-operation from stakeholders if it must work out for their benefit [64]. Research institutions will explore new algorithms and methodologies, and industry partnerships will help to bring these innovative solutions to the market [65]. Government bodies will ensure that ethical and safety standards are met while balancing innovation with responsible regulation. In the long run, AI powered drones will change the way many industries look and work and create new avenues through which automation, precision, and efficiency will be achieved in different industries [66]. The regulators will have a key part to play in ensuring that these are used safely and responsibly, resulting in a balanced approach to innovation and societal impact [67–69].

7.8 CONCLUSION

The advent of drones and smart technology, will herald the age of innovation in a transforming manner across many sectors. The increased interlocking of drones with smart technologies such as artificial intelligence (AI), machine learning, and advanced sensors will lead to transforming industries such as agriculture, logistics, healthcare, surveillance, and environmental monitoring. This helps them work through intricate functions alone, enhancing the accuracy of collected data, the operational efficiency, and reduced involvement by humans. The synergy of drones with smart technologies shall make their application go from automatic delivery mechanisms to live disaster response applications to achieve economic and social prosperity. With this trend continuing, we are going to observe highly sophisticated systems with upgraded features, including autonomy, swarm intelligence, and edge computing, expanding tasks the drone can execute. This raises even bigger challenges: more sophisticated issues related to the rules governing these devices, ensuring confidentiality over the data obtained and how this is to be further technologically advanced in various fronts. These would have to be overcome for drones and smart technologies to really make their mark on the future of diverse industries.

REFERENCES

[1] Heermann, P.D. (2015). Unmanned aerial systems (No. SAND2015-9558PE). Sandia National Lab. (SNL-NM), Albuquerque, NM (United States).

[2] Cavoukian, A. (2012). *Privacy and Drones: Unmanned Aerial Vehicles* (pp. 1–30). Information and Privacy Commissioner of Ontario.

[3] Colomina, I., Molina, P. (2014). Unmanned aerial systems for photogrammetry and remote sensing: a review. *ISPRS J. Photogrammetry Remote Sens.* 92, 79–97.

[4] FAA draft advisory circular: Unmanned air vehicle design criteria, Section 6.j, 1994, July 15.

[5] https://www.researchandmarkets.com/reports/4455437/global-unmanned-aerial-vehicle-uavmarket-value (accessed on 5November2018).

[6] https://www.businessinsider.com/commercial-uav-market-analysis-2017-8 (accessed on 5November2018).

[7] Ray, P.P. (2018). A survey on Internet of Things architectures. *J. King Saud Univ. Comput. Inf. Sci.*, 30(3), 291–319.

[8] Menkhoff, T., Kan, S.N., Tan, E.K., Foong, S. (2022). Future-proofing students in higher education with unmanned aerial vehicles technology: A knowledge management case study. *Knowledge Management & E-Learning: An International Journal*, 14(2), 223.
[9] Dai, M., Huang, N., Wu, Y., Gao, J., Su, Z. (2022). Unmanned-aerial-vehicle-assisted wireless networks: Advancements, challenges, and solutions. *IEEE Internet of Things Journal*, 10(5), 4117–4147.
[10] Mohsan, S.A.H., Khan, M.A., Noor, F., Ullah, I., Alsharif, M.H. (2022). Towards the unmanned aerial vehicles (UAVs): A comprehensive review. *Drones*, 6(6), 147.
[11] Emimi, M., Khaleel, M., & Alkrash, A. (2023). The current opportunities and challenges in drone technology. In *2024 4th International Conference on Pervasive Computing and Social Networking (ICPCSN)*. https://doi.org/10.1109/ICPCSN62568.2024.00106
[12] Maghazei, O., Steinmann, M. (2020). Drones in railways: Exploring current applications and future scenarios based on action research. *European Journal of Transport and Infrastructure Research*, 20(3), 87–102.
[13] Saeed, R.A., Omri, M., Abdel-Khalek, S., Ali, E.S., Alotaibi, M.F. (2022). Optimal path planning for drones based on swarm intelligence algorithm. *Neural Computing and Applications*, 34(12), 10133–10155.
[14] Ullah, Z., Al-Turjman, F., Mostarda, L., Gagliardi, R. (2020). Applications of artificial intelligence and machine learning in smart cities. *Computer Communications*, 154, 313–323.
[15] Anandaraj, A.P.S., Dhivya, R., Ateeq, K., Subramaniam, S. (2024). A survey of AI integration in unmanned aerial vehicles (UAVs) using digital twin technology: Advancements and applications. In *Digital Twin Technology and AI Implementations in Future-Focused Businesses* (pp. 14–26). IGI Global.
[16] Osco, L.P., Junior, J.M., Ramos, A.P.M., de Castro Jorge, L.A., Fatholahi, S.N., de Andrade Silva, J., Matsubara, E.T., Pistori, H., Gonçalves, W.N., Li, J. (2021). A review on deep learning in UAV remote sensing. *International Journal of Applied Earth Observation and Geoinformation*, 102, 102456.
[17] Amazon Prime Air's First Customer Delivery. Available online: https://www.youtube.com/watch?v=vNySOrI2Ny8&t=9s (accessed on 2January2018).
[18] Margaritoff, M. Drones in agriculture: How UAVs farming more efficient. Available online: http://www.thedrive.com/tech/18456/drones-in-agriculture-how-uavs-make-farming-more-efficient (accessed on 4February2018).
[19] Branson-Potts, H.L.A. Fire department used drones for the first time during skirball fire. Available online: http://www.latimes.com/local/lanow/la-me-ln-lafd-drone-skirball-fire-20171214-story.html (accessed on 11January2018).
[20] Walsh, K. (2020). 7 fun ways teachers can use drones for teaching and learning. Available online: http://www.emergingedtech.com/2017/10/7-fun-ways-teachers-can-used-drones-for-teaching-and-learning (accessed on 12January2018).
[21] Cenejac, J. (2021). 5 ways to use drones in the classroom: Cherishing students' passion for technology. Available online: http://elearningindustry.com/drones-in-the-classroom-5-ways-cherishing-studentspassion-technology (accessed on 11January2018).
[22] Carnahan, C., Zieger, L., Crowley, K. (2016). *Drones in education: Let your students' imagination soar*. International Society for Technology in Education.

[23] Smith, B., Mader, J. (2018). Drones for the science classroom. *Sci. Teach.*, 85, 16.

[24] Hoffert, F. Examining the possibility of using programming language with low-cost drones. In *Proceedings of the 2017 International Conference on Applied Computer and Communication Technologies*, Jakarta, Indonesia, 17–18 May 2017.

[25] Hussey, M. Drones in woodland elementary classroom soar, flip and teach. Available online: http://www.tampabay.com/news/education/k12/drones-in-woodland-elementary-classroom-soar-flipand-teach/2260352 (accessed on 15January2018).

[26] Wakefield, J. Robots and drones take over classrooms. Available online: http://www.bbc.com/news/technology-38758980 (accessed on 16January2018).

[27] Schaeffer, D.M., Olson, P.C. (2017). Drones in the classroom. *J. Comput. Sci. Coll.*, 32, 85–91.

[28] Carnahan, C., Crowley, K. (2018, March 5). Using drones to ensure student success. In *Proceedings of the 2017 Society for Information Technology & Teacher Education International Conference*, Austin, TX, USA.

[29] Palaigeorgiou, G.; Malandrakis, G.; Tsolopani, C. Learning with drones: Flying windows for classroom virtual field trips. In *Proceedings of the 2017 IEEE 17th International Conference on Advanced Learning Technologies*, Timisoara, Romania, 3–7 July 2017.

[30] Gerson, B.P., Sorby, S.A., Wysocki, A., Baartmans, B.J. (2001). The development and assessment of multimedia software for improving 3D spatial visualization skills. *Comput. Appl. Eng. Educ.*, 9, 105–113.

[31] Sorbi, S.A. (1999). Developing 3D spatial visualization skills. *Eng. Des. Gr. J.*, 63, 21–32.

[32] Sorbi, S.A. (2009). Educational research in developing 3D spatial skills for engineering students. *Int. J. Sci. Educ.*, 31, 459–480.

[33] Baki, A., Kosa, T., Guven, B. (2011). A comparative study of the effects of using dynamic geometry software and physical manipulative on the spatial visualization skills of preservice mathematics teachers. *Br. J. Educ. Technol.*, 42, 291–310.

[34] Chou, P.N., Chen, W.F., Wu, C.Y., Carey, R.P. (2017). Utilizing 3D open source software to facilitate student learning of fundamental engineering knowledge: A quasi-experimental study. *Int. J. Eng. Educ.*, 33, 382–388.

[35] Fombuena, A. (2017). Unmanned aerial vehicles and spatial thinking: Boarding education with geotechnology and drones. *IEEE Geosci. Remote Sens.*, 5, 8–18.

[36] Kazakoff, E.R., Sullivan, A., Bers, M.U. (2013). The effect of a classroom-based intensive robotics and programming workshop on sequencing ability in early childhood. *Early Child. Educ. J.*, 41, 245–255.

[37] Massachusetts Department of Elementary and Secondary Education. Kindergarten learning experience. Available online: https://www.doe.mass.edu/frameworks/0408kle.docx (accessed on 21January2018).

[38] Sarama, J., Clements, D.H. (2003). Building blocks of early childhood mathematics. *Teach. Child. Math.*, 9, 480–484.

[39] Zelazo, P.D., Carter, A., Reznick, J.S., Frye, D. (1997). Early development of executive function: A problem-solving framework. *Rev. Gen. Psychol*, 1, 198–226.

[40] Fessakis, G., Gouli, E., Mavroudi, E. (2013). Problem solving by 5–6 years old kindergarten children in a computer programming environment: A case study. *Comput. Educ.*, 63, 87–97.
[41] Saez-Lopez, J.M.; Roman-Gonzalez, M.; Vazquez-Cano, E. (2016). Visual programming languages integrated across the curriculum in elementary school: A two year case study using scratch in five schools. *Comput. Educ.*, 97, 129–141.
[42] Brennan, K., & Resnick, M. (2012). New frameworks for studying and assessing the development of computational thinking. In *Annual American Educational Research Association Meeting*. American Educational Research Association.
[43] Teaching and learning for sustainable future. Available online: http://www.unesco.org/education/tlsf/ (accessed on 1October2018).
[44] ISTE standards for students. Available online: https://www.iste.org/standards/for-students (accessed on 1October2018).
[45] Feruglio, L., Benetton, A., Varile, M., Vittori, D., Bloise, I., Maderna, R., Cardenio, C., Madonia, P., Rossi, F., Azza, F.P., & De Marchi, P. (2023). Future-ready space missions enabled by end-to-end AI adoption. In *Artificial Intelligence for Space: AI4SPACE* (pp. 303–360). CRC Press.
[46] Dayo-Olupona, O., Genc, B., Celik, T., & Bada, S. (2023). Adoptable approaches to predictive maintenance in the mining industry: An overview. *Resources Policy*, 86, 104291.
[47] Latapie, H., Gabriel, M., Srinivasan, S., Kompella, R., Thórisson, K.R., & Wang, P. (2023, September). Automated spatiotemporal modeling for real-time data-driven actionable insights. In *Intelligent Systems Conference* (pp. 780–798). Springer Nature Switzerland.
[48] Hall, O., & Wahab, I. (2021). The use of drones in the spatial social sciences. *Drones*, 5(4), 112.
[49] Mandal, A., & Ghosh, A.R. (2023). Role of artificial intelligence (AI) in fish growth and health status monitoring: A review on sustainable aquaculture. *Aquaculture International*, 32, 2791–2820.
[50] Zakari, R.Y., Shafik, W., Kalinaki, K., & Iheaturu, C.J. (2024). 15 Internet Things (IoFT) of Forestry technologies and applications in forest management. *Advanced IoT Technologies and Applications in the Industry 4.0 Digital Economy* (p. 275). CRC Press.
[51] Nguyen, D.D., Rohacs, J., & Rohacs, D. (2021). Autonomous flight trajectory control system for drones in smart city traffic management. *ISPRS International Journal of Geo-Information*, 10(5), 338.
[52] Akindote, O.J. (2023, May). Digital Era: navigating VMI and supplychain for sustainable success. In *CS & IT Conference Proceedings* (Vol. 13, No. 9). CS & IT Conference Proceedings.
[53] Babarinde, A.O., Ayo-Farai, O., Maduka, C.P., Okongwu, C.C., & Sodamade, O. (2023). Data analytics in public health, A USA perspective: A review. *World Journal of Advanced Research and Reviews*, 20(03), 211–224.
[54] Ogundairo, O., Ayo-Farai, O., Maduka, C.P., Okongwu, C.C., Babarinde, A.O., & Sodamade, O.T. (2023). Review on Maldi mass spectrometry and its application in clinical research. *International Medical Science Research Journal*, 3(3), 108–126.
[55] Orieno, O.H., Ndubuisi, N.L., Ilojianya, V.I., Biu, P.W., & Odonkor, B. (2024). The future of autonomous vehicles in The US urban landscape: A review:

Analyzing implications for traffic, urban planning, and the environment. *Engineering Science & Technology Journal*, 5(1), 43–64.

[56] Lu, B., Dao, P.D., Liu, J., He, Y., & Shang, J. (2020). Recent advances of hyperspectral imaging technology and applications in agriculture. *Remote Sensing*, 12(16), 2659.

[57] Ohenhen, P.E., Chidolue, O., Umoh, A.A., Ngozichukwu, B., Fafure, A.V., Ilojianya, V.I., & Ibekwe, K.I. (2024). Sustainable cooling solutions for electronics: A comprehensive review: Investigating the latest techniques and materials, their effectiveness in mechanical applications, and associated environmental benefits. *World Journal of Advanced Research and Reviews*, 21(1), 957–972.

[58] Babatunde, F.O., Omotayo, A.B., Oluwole, O.I., & Ukoba, K. (2021, April). A review on wastewood reinforced polymer matrix composites for sustainable development. In *IOP Conference Series: Materials Science and Engineering* (Vol. 1107, No. 1, p. 012057). IOP Publishing.

[59] Lukong, V.T., Ukoba, K., Yoro, K.O., & Jen, T.C. (2022). Annealing temperature variation and its influence on the self-cleaning properties of TiO2 thin films. *Heliyon*, 8(5).

[60] Illiashenko, O., Kharchenko, V., Babeshko, I., Fesenko, H., & Di Giandomenico, F. (2023). Security-Informed safety analysis of autonomous transport systems considering AI-powered cyberattacks and protection. *Entropy*, 25(8), 1123.

[61] Miya, J., Raj, S., Ansari, M.A., Kumar, S., & Kumar, R. (2023). Artificial intelligence advancement for 6G communication: A visionary approach. In *6G Enabled Fog Computing in IoT: Applications and opportunities* (pp. 355–394). Springer Nature Switzerland.

[62] Fascista, A. (2022). Toward integrated large-scale environmental monitoring using WSN/UAV/Crowdsensing: A review of applications, signal processing, and future perspectives. *Sensors*, 22(5), 1824.

[63] Habbal, A., Ali, M.K., & Abuzaraida, M.A. (2024). Artificial intelligence trust, risk and security management (AI TRISM): Frameworks, applications, challenges and future research directions. *Expert Systems with Applications*, 240, 122442.

[64] Maghazei, O., & Netland, T. (2020). Drones in manufacturing: Exploring opportunities for research and practice. *Journal of Manufacturing Technology Management*, 31(6), 1237–1259.

[65] Bayomi, N., & Fernandez, J.E. (2023). Eyes in the sky: drones' applications in the built environment under climate change challenges. *Drones*, 7(10), 637.

[66] Sharma, S. (2023). Trustworthy artificial intelligence: design of AI governance framework. *Strategic Analysis*, 1–22.

[67] Layton, P. (2021). Fighting artificial intelligence battles: Operational concepts for future AI enabled wars. *Network*, 4(20), 1–100.

[68] Ezeigweneme, C.A., Umoh, A.A., Ilojianya, V.I., & Adegbite, A.O. (2024). Review of telecommunication regulation and policy: Comparative analysis USA and Africa. *Computer Science & IT Research Journal*, 5(1), 81–99.

[69] Salas, B., Salcedo, R., Garcia-Ruiz, F., & Gil, E. (2023). Design, implementation and validation of a sensor-based precise airblast sprayer to improve pesticide applications in orchards. *Precision Agriculture*, 25, 865–888.

Chapter 8

Potential applications, challenges and future directions of artificial intelligence in healthcare

A review

Mukul Kumar Gupta and Abhinav Sharma

8.1 INTRODUCTION

Artificial intelligence (AI) is the simulation of human intellect in devices that are programmed to mimic the human behavior [1]. AI-based machines can perform decision-making tasks without requiring human interpretation [2]. This interdisciplinary approach can be applied to various fields, including medicine and health. Initially, research studies attempted to enhance diagnosis methods using computer-aided algorithms, and ever since, AI has found wide applications in the healthcare sector. In recent years, advancements in computational capacity and the availability of vast medical data have significantly increased the use of AI in medical applications [3–6]. AI influences medical practice in many ways, including therapeutic, diagnostic, rehabilitative, surgical, and prognostic therapies [7]. AI is also helpful in two critical areas of medicine: clinical decision-making and disease diagnosis. AI systems can collect, evaluate, and report vast data from many modalities to identify diseases and assist healthcare decisions [8]. These systems are powerful in handling large amounts of medical data created in the medical profession and extracting relevant information from big medical data [3]. In some studies, researchers have used AI to discover and develop new drugs [9]. The use of AI is bold, according to an assessment of significant research databases. AI significantly reduces care expenses and repetitive operations and allows medical professionals to work on critical thinking and clinical innovation [10]. Some studies point out that the AI perspective is promising; nonetheless, there is scope for more research to accomplish the efficacy and find applications of AI in the medical field [10–12]. Figure 8.1 shows the prominent healthcare applications of AI in our day-to-day life. AI usually refers to a collection of software and hardware components. AI is mainly concerned with algorithms from a software perspective.

A theoretical foundation for performing AI algorithms is an artificial neural network (ANN). It's a computational model based on the human brain, consisting of a network of linked neurons connected by weighted communication channels [13]. A single neuron responds to several inputs from nearby neurons, and the network varies its state due to external inputs [13].

DOI: 10.1201/9781003568995-8

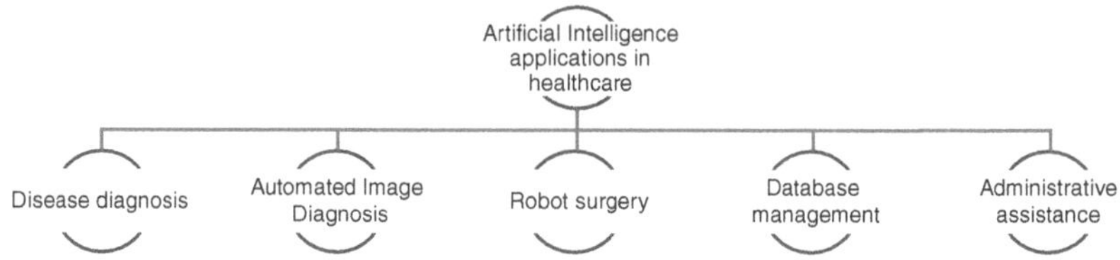

Figure 8.1 Major applications of AI in healthcare.

Therefore, neural network (NN) responds to environmental inputs, like how the human brain responds to various environmental inputs. NNs are layered structures that can be configured in a variety of ways. NNs can perform tasks by either using supervised learning or unsupervised learning. The former refers to the task where the function maps an input to an output based on a labeled dataset while the latter refers to the task where it learns from the data which is not labeled. In unsupervised learning, models classify or categorize by identifying common features in the data. As processing power has increased, NNs have become "deeper" and with more computational power NNs can exploit deeper network to replicate a human [14].

From a hardware aspect, AI is primarily concerned with implementing NN algorithms on a high-performance computer. The simplest method uses a multithread or multicore general-purpose central processing unit (CPU) to run the NN algorithm [15]. Furthermore, graphical processing units (GPUs), which excel at convolutional calculations, outperform CPUs for large-scale NNs [16]. Particularly for spiking NNs, co-processing with the CPU and GPU has proven to be more efficient than processing with the CPU alone [16]. Additionally, more advanced systems, namely, field-programmable gate arrays (FPGAs) and application-specific integrated circuits (ASICs), are programmable or customizable accelerator hardware platforms [17]. In most medical fields, like CT (computed tomography) scanning and MRI (magnetic resonance imaging), fast and precise results are required from complicated algorithms. CT and MRI equipment manufacturers have a major problem cutting computation time down. In some specific medical applications, different kinds of calculation and visualization skills are required. To give just one illustration, high-fidelity displays are essential to various technology solutions that enable medical professionals to view and assess complicated pictures [18]. The medical field has benefited from the high computing systems and AI technologies.

Recent advancements in AI software and hardware technologies have led to wide applications in many technological sectors, including the Internet of Things (IoT), machine vision, autonomous driving, natural language processing, and robotics [19–20]. Most intriguingly, biomedical researchers have been actively attempting to use AI to enhance analysis and treatment outcomes, increasing the healthcare sector's efficacy. AI and its applications in biology have lately achieved great strides over the last decade. This chapter reviews key biomedical engineering and healthcare topics that are benefited by current AI technological developments. Healthcare should be more individualized, predictive, preventive, and interactive, and AI shows the potential to advance these goals. It has been predicted that AI will maintain its momentum and will develop and mature as a potent tool for biomedical applications. The organization of this chapter is as follows. Section 8.2 focuses on prominent AI tools used in healthcare. Section 8.3 describes healthcare applications of AI, IoT, and robotics. Section 8.4 reports the future scope of AI and IoT and section 8.5 provides a conclusive remark to summarize the chapter.

8.2 PROMINENT AI TOOLS IN HEALTHCARE

AI has immense potential to improve people's health and boost healthcare delivery efficiency. The leading AI technologies are machine learning, neural networks, and reinforcement learning. Although the procedures and jobs that these technologies serve are various, they are all directly pertinent to healthcare. The following sections identify and explain some of healthcare's most important AI technologies.

8.2.1 Machine learning

Machine Learning (ML) is a sub-domian of AI and produces data analysis algorithms to extract meaningful information from data. In medical applications, patient 'traits' and, occasionally, medical outcomes of interest are the input parameters of ML algorithms. A patient's traits typically include baseline data such as age, gender, sickness history and so on. Individuals' medical outcomes, as well as their characteristics, are commonly collected in clinical research. Among them are illness markers, patient survival times, and quantitative disease levels such as tumour sizes [21].

Unsupervised and supervised learning are two types of ML algorithms that differ in whether the data is labeled or not unlabeled as shown in Figure 8.2. Unsupervised learning is used for extracting features, whereas supervised learning is well-suited for prediction by establishing correlations between patient features (as input) and the intended outcome (as output) [22]. Semi-supervised learning is a combination of unsupervised and supervised learning. It is in situations when the outcomes of some patients are unknown.

Clustering and principal component analysis (PCA) are common unsupervised learning techniques. Clustering involves grouping data points with similar characteristics into clusters without relying on predefined outcome labels. In patient analysis, cluster labels are assigned using methods that enhance similarity within clusters while minimizing similarity between them. Well-known clustering techniques include K-means clustering, hierarchical

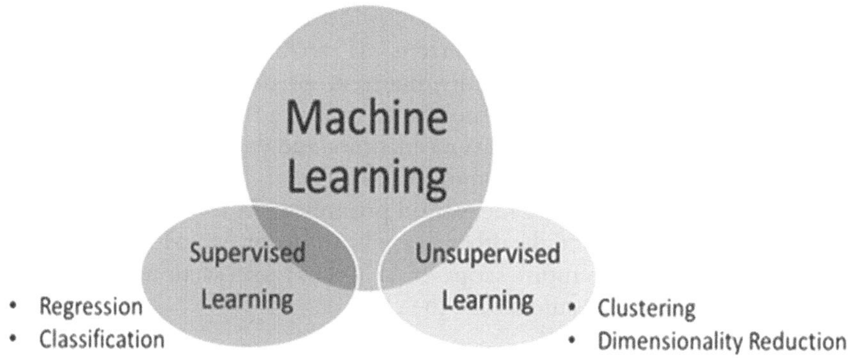

Figure 8.2 Types of machine-learning algorithms.

clustering, and Gaussian mixture models. On the other hand, PCA is mainly used for dimensionality reduction, especially in high-dimensional datasets, such as genome-wide association studies involving a large number of genes. [23]. PCA projects the data on a plane with principal components without losing too much information about the subjects [23].

On the other hand, supervised learning involves training a computational model on input data labeled for certain outputs. The model formulations, in most cases, differ based on the desired outcomes. For example, outcomes can include the likelihood of encountering a given clinical event, the expected value of a disease level, or the expected survival duration. Unlike unsupervised learning, supervised learning is employed most frequently in AI healthcare applications. In practice, researchers use unsupervised learning in the preprocessing step to reduce dimensionality, which makes the subsequent supervised learning step more efficient [24].

8.2.2 Neural networks

Neural networks composed of parallel-processing computing units referred to as neurons and is inspired from biological network of neurons. [13]. Neurons are the brain's fundamental unit, which collect information from the environment and analyze it before sending electrical impulses to the body for a reaction. Dendrites, nucleus, cell body, axons, and synapses are the fundamental components of biological neurons. Dendrites receive messages from other neurons to which they are attached, and the messages are subsequently sent along axons to other neurons' dendrites (there exists a junction between the dendrites and axons known as synapses). An artificial neuron's structure resembles biological neurons in the brain in terms of connection. ANN is composed of different layers of neurons. Neurons in the first layer communicate via channels with neurons in the next layer. Inputs for each neuron is multiplied by synaptic weight. The information from active neurons is transferred as inputs to the next layer, and this process is continued until the output is produced. Thus, we train the neural network model through repeated forward and backward propagation.

8.2.3 Deep learning

Deep learning is the subset of traditional ML algorithms and allows computational models composed of many processing layers based on neural networks to learn multiple degrees of abstraction for data representations [14].

The primary differences between deep learning and traditional ANNs are the number of hidden layers, their connections, and the ability to get meaningful abstractions of the inputs. Contrarily, traditional ANNs are frequently restricted to three layers and trained to provide supervised representations that are only seldom generalizable and best suited to the task at hand. Alternately, each layer of a deep learning system optimizes a local unsupervised criteria to construct a representation of the observed patterns based on the data it gets as inputs from the preceding layers.

Furthermore, there has been a breakthrough in unsupervised pre-training [25] and new methods to prevent overfitting [26] have been developed. General-purpose graphics processing units to speed up computations, and the development of high-level modules to easily build NN have been established. Deep learning models provide state-of-the-art solutions for a variety of tasks (e.g. Theano [27], Caffe [28], TensorFlow [14]). With remarkable achievements in image object detection, audio recognition, natural language understanding, deep learning has proven to be useful in detecting precise patterns in high-dimensional data [29].

8.3 APPLICATIONS OF AI, IOT, AND ROBOTICS IN HEALTHCARE

8.3.1 Diagnostics and prognosis of diseases using AI

Despite the rising use of ML in healthcare, the areas that have received the most attention include neurological systems, cancer, and heart disorders, which are the leading causes of death and disability. Chronic and infectious disorders, such as inflammatory bowel disease and type 2 diabetes, have received much attention in recent years. By improving our understanding of clinical data and leveraging it to train models and create diagnosis systems, early diagnosis for diseases can be achieved [1].

The model detected benign instances and recognized malignant cases in a study of 1,634 malignant and benign lung tissue images. According to research, depression affects about 6% of the adult population in the United States. Image heatmap pattern recognition was shown to be 74% accurate in diagnosing depression [30].

For cardiac abnormalities, cloud-based ANN algorithms such as Cardio DL and Arteries are utilized to diagnose cardiac irregularities utilizing medical imaging data, in this case, MRI images of the heart ventricles [31]. In comparison to radiologists, the instruments were proven to analyze heart ventricular and blood flow function [32]. According to the latest estimates, Chronic Kidney Condition (CKD) is a dangerous disease that affects 2.5-11.2 percent of individuals in Australia, Asia, Europe, and North America. The AI-based hybrid model has a 97.8% accuracy rate in predicting CKD [33].

8.3.2 Healthcare IoT

Wearable sensor devices are used to track human health and activity. Especially nowadays smart watches and apps have been developed through which we can monitor heart rate or the amount of water we are consuming. As shown in Figure 8.3, activity trackers with IoT assist in problems like obesity and insomnia. In general, activity trackers have been part of a "next generation" of wearable computers that transformed how people live and function in the twenty-first century. Figure 8.3 shows real-life examples of how IoT improves every life, especially related to healthcare. One can check various parameters like heart rate, step count, calorie count, and moving steps in just one click. There are three main layers in IoT hardware as described in Figure 8.4.

i. **Perception layer**: Perception layer is attended to the hardware layer, and it's main role is to collect the data and then prepare it to send it to the network layer, where the network layer connects all smart devices and data exchange happens via this.

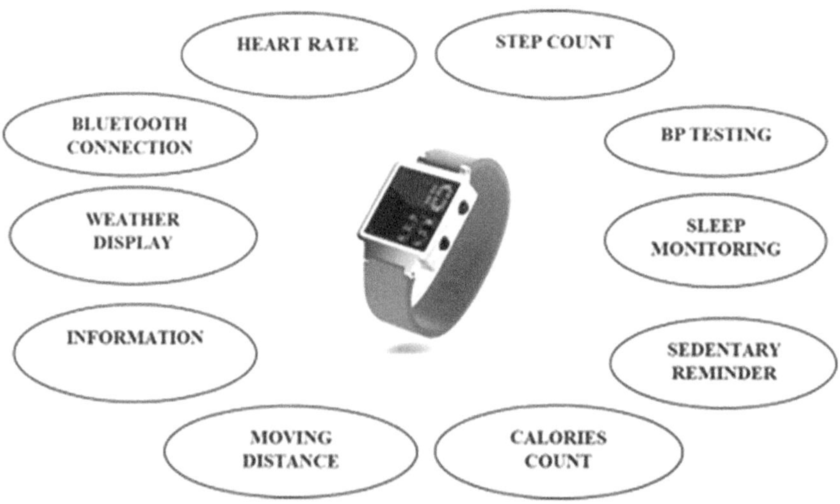

Figure 8.3 Activity tracking using IoT.

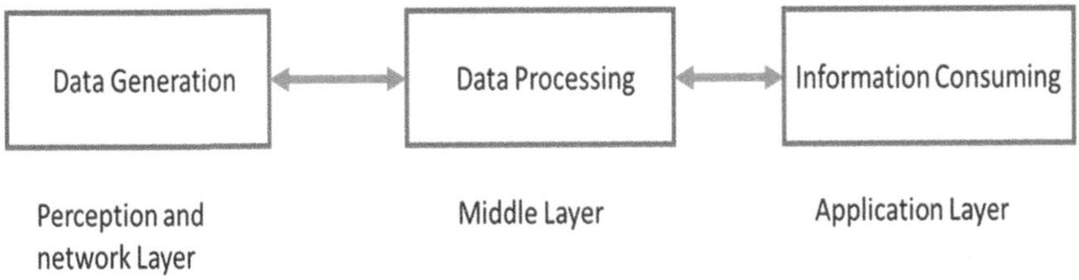

Figure 8.4 IoT based hardware.

ii. **Middleware layer:** It stores the health data in a database.
iii. **Application Layer:** It provides healthcare-related applications and services to satisfy the endpoint customer.

Further, there are three main steps in healthcare IoT systems: data generation, data processing, and information consumption.

(i) Data generation: Collect and generate the needed data through various devices and even gather data from patients. This is performed through the perception and network layer.
(ii) Data processing: It is analyzed through a well-established ML technique which is implemented in the middleware layer.
(iii) Information consuming: The outcomes and analytics of the data processing can be done in this phase, and the decision has been performed.

The IoT permits devices to exchange information and carry out tasks such as sensing, identifying, and networking. IoT is used for remote monitoring and controlling devices. Various sensors for measuring the parameters are placed on the hardware, and the data measured from these sensors are given to the microcontroller. In the Figure 8.5 IoT based hardware diagram has been shown where hardware is connected to the microcontroller with IoT, which is connected to a router and finally to the internet. In the current scenario, mainly wireless technology is reflected as it can cover a wide network area.

Monitoring the parameters inside a hospital can be prepared with a group of sensors that sense the information based on the detecting parameters. Each sensor in the group will perform a devoted task and will communicate straightforwardly with the controller. For instance, one sensor may be used to check pulses, whereas the other sensor may be used to identify blood level.

8.3.3 Healthcare robotics

Robots nowadays are mostly used in making the life of humans easy rather than conventional automation. Human friendly means robots are helping in medical related cases like robotic assisted surgery, lower and upper limb

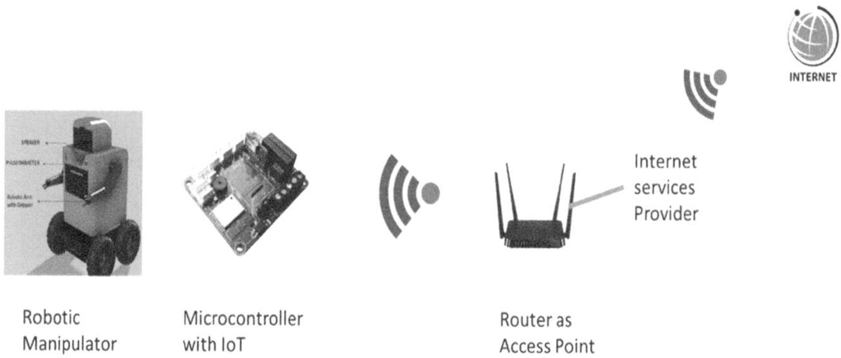

Figure 8.5 IoT Healthcare architecture.

amputees, robotic instruments like medical imaging, tumor detection all can be done easily with the help of advanced robotics. The advantage of using robotics is that most of them are noninvasive in nature, painless treatment, more accurate and precise in nature. Robots are also useful in mental therapy, personal assistance, brain computer interface where a human brain and external device to talk to one another – to exchange signals.

Major advantages of integrating IoT and robotics in healthcare are (i) reduction in treatment costs to be incurred by the patient and (ii) comparatively fast and more efficient results.

Improving healthcare is one of the most important concerns across the world nowadays. The cost of healthcare is increasing exponentially due to various factors like shortage of trained doctors, increasing infrastructure costs, medical equipment costs, etc., which finally leads to the burden on every individual.

In healthcare, healthcare monitoring is one of the major areas that need to be addressed. In healthcare monitoring, Human-Machine interaction (HMI) is an important aspect. In HMI, if any device or hardware needs to be integrated, then there should not be any time delay, and the system should have a proper response so that all the data can be taken without delay. The idea is intended to design a health-monitoring manipulator so that it can be beneficial to society. The elderly population is increasing daily with a labor shortage, resulting in an ultimate challenge to society. Apart from the assistive social robot, the main application will be the health monitoring system. The CAD model of four Wheeled Mobile Robots (WMR) is shown in Figure 8.6 where the main aim is to design and develop a manipulator which can be used mainly for the health monitoring system. The focus is to develop an assistive social robot with intelligence that can enable people to live independently.

The model structure contains various health monitoring sensor like pulse oximeter, blood pressure measuring sensor, heart pulse measuring sensor,

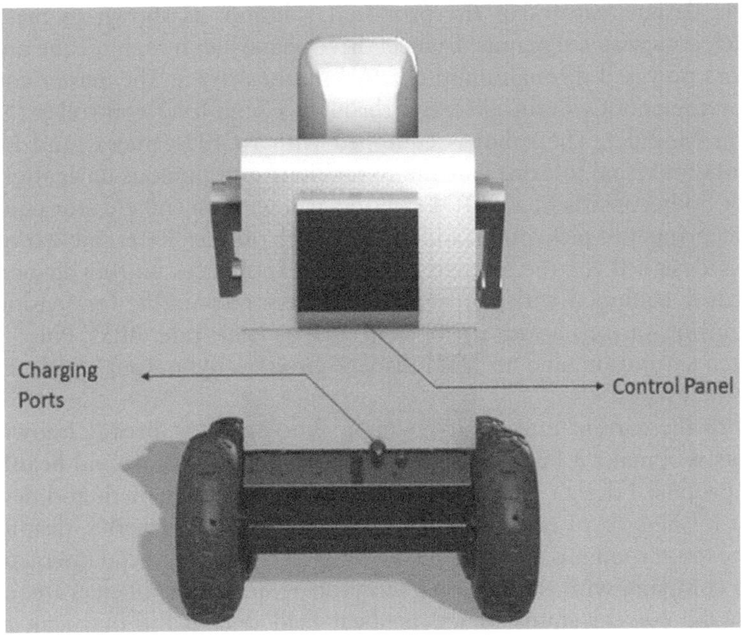

Figure 8.6 CAD model of WMR for healthcare assistance.

bio sensor, oximeter sensor, glucose sensor, temperature sensor, electromyogram sensor, other health related sensor can be added based on requirement so that health related data can be gathered for data analysis and monitoring.

Figure 8.7 shows the basic diagram, which is a closed-loop control system with feedback from the sensor to update the state variables for improving the robot navigation. The system consists of a high-level controller (Nvidia Jetson Module) driven by a multichannel motor driver to power up the motors and other actuators to move the robot. The sensor information from the Lidar and visual feedback from the cameras help navigate the robot to the desired location (waypoints), providing feedback on the robot's position with respect to its surroundings. The Bio signal sensor provides sensor output to the HMI display and sensor feedback for arm actuation.

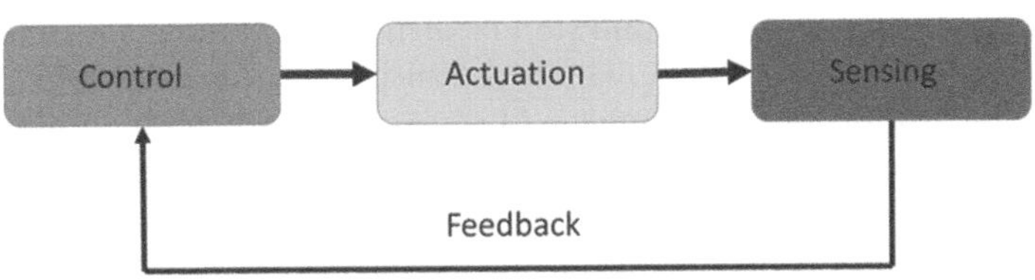

Figure 8.7 Block diagram of closed-loop control system with feedback.

The design features of the proposed solution, as shown in the robot model, comprise a dynamic 4-wheel drive chassis with high torque and DC motors powered through multi-channel motor drivers. The master controller or the robot's brain is served through a high-level controller (Nvidia Jetson Module). The robot is equipped with RGBD cameras and LIDAR sensors for visual information and to facilitate autonomous navigation. The robot body consists of a 3 DOF robotic arm with an end effector equipped with a gripper to pick, place, and interact with objects. Retractable tray storage is designed for the storage of medical appliances and medicines. The system is equipped with bio-signaling sensors responsible for sensing and diagnosing various health parameters such as heart rate, BPM, Pulse, ECG, oxygen saturation, and an HMI display provides the output and facilitates user interaction.

With the current rate of advancing technology, it is already known that robots will make a huge difference in the field of medicine and healthcare. The proposed design is about to offer a great deal of growth and development in the area of healthcare robots by assisting in surgeries, disinfecting the hospital rooms, dispensing medication, and offering social interaction to keep company with the patients, monitoring health parameters frequently. With the current status of the medical field due to the outbreak of the COVID19 virus, hospitals shall benefit from using healthcare robots to assist them in performing necessary formalities and tests. This model will be helpful for guidance, physical support, health monitoring, medicine, and other scheduling. Robotics plays an important role in health monitoring and personal aids for mobility with the help of current technologies.

Robotics in Surgery: The integration of healthcare and robotics is profoundly connected in the field of surgery, significantly assisting the surgeon's hand. The significant surgical care by robots includes simulation training, intraoperative navigation, postoperative assessment, open surgery, minimally invasive procedures, and remote surgery. It reduces surgical trauma, diminishes tissue damage, improves the patient's recovery period, appreciates the operation without a tremor, and favors precise micro-motions aided with micro-forces.

8.3.4 Management of medical data using AI

The most apparent use of AI in healthcare is data management. Collection of data from various sensors and standardizing it, and tracing its lineage are all things that need to be done. This is the initial step toward overhauling the current healthcare system. Google's AI research branch, has announced that the Google Deep Mind Health initiative, which analyzes medical facts to give excellent and timely health services [34]. Since acquiring and evaluating data is a crucial step, data management is the most extensively used application of AI in health care. Robots acquire, store, re-layout, and track data to allow faster, more consistent access [20]. The amount of health information available has increased during the last ten years. Massive volumes of data such as patient diagnostic information, novel research findings are produced every day in the medical sector [3]. The utilization of big data and innovative tools helps organizations to communicate more effectively with patients and make better decisions [35]. The application of big data has grown substantially in recent years, from cost reduction to optimizing hospital staff scheduling, from remote patient monitoring to anticipating epidemics.

8.4 FUTURE SCOPE

8.4.1 AI in healthcare

AI is a branch of computer science and technology that deals with developing intelligent behavior in machines. Physicians employ AI technologies with their own experience, knowledge, and human connection to provide high-quality patient care while cutting costs. AI can analyze big medical data gathered from large patient populations, uncover new evidence, and identify high-quality healthcare facilities [10].

Seeing the vast potential of AI, researchers feel that healthcare organizations should spend on AI technology that would be advantageous in the future. Governments should invest in and contribute to modernization since these technologies are vital for managing health services and patient treatment. Furthermore, this should occur in the world's poorest countries, where infrastructure and services related to health and medication are in limited supply.

In the near future medical practitioners will explore deep-learning architectures such as convolution neural network (CNN) for diagnosing cancer and other diseases through ultrasound, CT scan and MRI medical images. AI built chatbots can also provide medical prescription to patients and make the life of humans easier, especially in remote areas where medical facility is unavailable.

8.4.2 IoT in healthcare

On the other hand, IoT devices are becoming more innovative as cloud services and AI become more widely used. They can now contribute more than merely transferring data from a patient to a healthcare provider. For example, the smart glucose monitoring system and smart insulin pen are IoT devices that process the data using cloud services [36]. The glucose monitoring device collects glucose level data in real-time. Subsequently, the IoT interface uploads the collected data to a cloud service or a mobile app for analysis. Finally, the insulin pump injects the patient with the proper insulin dosage based on the analysis results. More applications may involve sophisticated nanny cams to watch older people. In addition to wearables and patient-specific interactions, healthcare companies will use IoT in facilities for inventory management [37]. By monitoring its mobility and general use, hospitals can identify possible equipment shortages better and who could have come into contact with the equipment.

8.5 CONCLUSION

The chapter presents the diverse application of innovative technologies such as AI and IoT in medical industry. The different applications that are outlined in the chapter are diagnostics and prognosis of diseases using AI, IoT architecture and devices for healthcare, robotics for healthcare specifically in surgery, and management of medical data. The authors also developed a robot model which can be useful for monitoring various healthcare parameters. With the increase in computing facilities, analysis of health-related data using AI algorithms can diagnose and control many diseases over the period. The chapter also discussed the future scope of AI and IoT in healthcare and how this technology can be useful for medical practitioners.

REFERENCES

[1] S. E. Dilsizian and E. L. Siegel, "Artificial intelligence in medicine and cardiac imaging: Harnessing big data and advanced computing to provide personalized medical diagnosis and treatment," *Curr Cardiol Rep*, vol. 16, no. 1, p. 441, Dec. 2013. https://doi.org/10.1007/s11886-013-0441-8

[2] T. Lysaght, H. Y. Lim, V. Xafis, and K. Y. Ngiam, "AI-assisted decision-making in healthcare," *ABR*, vol. 11, no. 3, pp. 299–314, Sept. 2019. https://doi.org/10.1007/s41649-019-00096-0

[3] T. B. Murdoch and A. S. Detsky, "The inevitable application of big data to health care," *JAMA*, vol. 309, no. 13, pp. 1351–1352, Apr. 2013. https://doi.org/10.1001/jama.2013.393

[4] N. S. Malan and S. Sharma, "Feature selection using regularized neighborhood component analysis to enhance the classification performance of motor imagery signals," *Computers in Biology and Medicine*, vol. 107, pp. 118–126, Apr. 2019. https://doi.org/10.1016/j.compbiomed.2019.02.009

[5] N. S. Malan and S. Sharma, "Motor imagery EEG spectral-spatial feature optimization using dual-tree complex wavelet and neighborhood component analysis," *IRBM*, Jan. 2021. https://doi.org/10.1016/j.irbm.2021.01.002

[6] N. Singh Malan and S. Sharma, "Time window and frequency band optimization using regularized neighborhood component analysis for multi-view motor imagery EEG classification," *Biomedical Signal Processing and Control*, vol. 67, p. 102550, May 2021. https://doi.org/10.1016/j.bspc.2021.102550

[7] S. Jha and E. J. Topol, "Adapting to artificial intelligence: Radiologists and pathologists as information specialists," *JAMA*, vol. 316, no. 22, pp. 2353–2354, Dec. 2016. https://doi.org/10.1001/jama.2016.17438

[8] S. Secinaro, D. Calandra, A. Secinaro, V. Muthurangu, and P. Biancone, "The role of artificial intelligence in healthcare: A structured literature review," *BMC Medical Informatics and Decision Making*, vol. 21, no. 1, p. 125, Apr. 2021. https://doi.org/10.1186/s12911-021-01488-9

[9] D. Paul, G. Sanap, S. Shenoy, D. Kalyane, K. Kalia, and R. K. Tekade, "Artificial intelligence in drug discovery and development," *Drug Discov Today*, vol. 26, no. 1, pp. 80–93, Jan. 2021. https://doi.org/10.1016/j.drudis.2020.10.010

[10] A. Bohr and K. Memarzadeh, "The rise of artificial intelligence in healthcare applications," *Artificial Intelligence in Healthcare*, pp. 25–60, 2020. https://doi.org/10.1016/B978-0-12-818438-7.00002-2

[11] D. D. Miller and E. W. Brown, "Artificial intelligence in medical practice: The question to the answer?" *Am J Med*, vol. 131, no. 2, pp. 129–133, Feb. 2018. https://doi.org/10.1016/j.amjmed.2017.10.035

[12] A. Lavecchia, "Deep learning in drug discovery: opportunities, challenges and future prospects," *Drug Discovery Today*, vol. 24, no. 10, pp. 2017–2032, Oct. 2019. https://doi.org/10.1016/j.drudis.2019.07.006

[13] O. I. Abiodun, A. Jantan, A. E. Omolara, K. V. Dada, N. A. Mohamed, and H. Arshad, "State-of-the-art in artificial neural network applications: A survey," *Heliyon*, vol. 4, no. 11, p. e00938, Nov. 2018. https://doi.org/10.1016/j.heliyon.2018.e00938

[14] R. Miotto, F. Wang, S. Wang, X. Jiang, and J. T. Dudley, "Deep learning for healthcare: review, opportunities and challenges," *Briefings in Bioinformatics*, vol. 19, no. 6, pp. 1236–1246, Nov. 2018. https://doi.org/10.1093/bib/bbx044

[15] J. Misra and I. Saha, "Artificial neural networks in hardware: A survey of two decades of progress," *Neurocomputing*, vol. 74, no. 1, pp. 239–255, Dec. 2010. https://doi.org/10.1016/j.neucom.2010.03.021

[16] K.-S. Oh and K. Jung, "GPU implementation of neural networks," *Pattern Recognition*, vol. 37, no. 6, pp. 1311–1314, Jun. 2004. https://doi.org/10.1016/j.patcog.2004.01.013

[17] S. Mittal, "A survey of FPGA-based accelerators for convolutional neural networks," *Neural Comput & Applic*, vol. 32, no. 4, pp. 1109–1139, Feb. 2020. https://doi.org/10.1007/s00521-018-3761-1

[18] P. A. Guze, "Using technology to meet the challenges of medical education," *Trans Am Clin Climatol Assoc*, vol. 126, pp. 260–270, 2015.

[19] S. Pinto, J. Cabral, and T. Gomes, "We-care: An IoT-based health care system for elderly people," in *2017 IEEE International Conference on Industrial Technology (ICIT)*, Mar. 2017, pp. 1378–1383. https://doi.org/10.1109/ICIT.2017.7915565

[20] A. Maibaum, A. Bischof, J. Hergesell, and B. Lipp, "A critique of robotics in health care," *AI & Soc*, vol. 37, no. 2, pp. 467–477, Jun. 2022. https://doi.org/10.1007/s00146-021-01206-z

[21] F. Jiang et al., "Artificial intelligence in healthcare: past, present and future," *Stroke Vasc Neurol*, vol. 2, no. 4, pp. 230–243, Jun. 2017. https://doi.org/10.1136/svn-2017-000101

[22] K. Shailaja, B. Seetharamulu, and M. A. Jabbar, "Machine learning in healthcare: A review," in *2018 Second International Conference on Electronics, Communication and Aerospace Technology (ICECA)*, Mar. 2018, pp. 910–914. https://doi.org/10.1109/ICECA.2018.8474918

[23] G. T. Reddy et al., "Analysis of dimensionality reduction techniques on big data," *IEEE Access*, vol. 8, pp. 54776–54788, 2020. https://doi.org/10.1109/ACCESS.2020.2980942

[24] L. Zhou, S. Pan, J. Wang, and A. V. Vasilakos, "Machine learning on big data: Opportunities and challenges," *Neurocomputing*, vol. 237, pp. 350–361, May 2017. https://doi.org/10.1016/j.neucom.2017.01.026

[25] S. Saba-Sadiya, E. Chantland, T. Alhanai, T. Liu, and M. M. Ghassemi, "Unsupervised EEG artifact detection and correction," *Frontiers in Digital Health*, vol. 2, 2021. Accessed: Jun.11, 2022. [Online]. Available: https://www.frontiersin.org/article/10.3389/fdgth.2020.608920

[26] R. Roelofs et al., "A meta-analysis of overfitting in machine learning," *Advances in Neural Information Processing Systems*, vol. s, 2019. Accessed: Jun.27, 2022. [Online]. Available: https://proceedings.neurips.cc/paper/2019/hash/ee39e503b6bedf0c98c388b7e8589aca-Abstract.html

[27] S. Bahrampour, N. Ramakrishnan, L. Schott, and M. Shah, "Comparative study of deep learning software frameworks." arXiv, Mar. 29, 2016. https://doi.org/10.48550/arXiv.1511.06435

[28] Y. Jia et al., "Caffe: Convolutional Architecture for Fast Feature Embedding," in *Proceedings of the 22nd ACM international conference on Multimedia*, New York, NY, USA, Nov. 2014, pp. 675–678. https://doi.org/10.1145/2647868.2654889

[29] A. Kaur, Y. Singh, N. Neeru, L. Kaur, and A. Singh, "A survey on deep learning approaches to medical images and a systematic look up into real-time object detection," *Arch Computat Methods Eng*, vol. 29, no. 4, pp. 2071–2111, Jun. 2022. https://doi.org/10.1007/s11831-021-09649-9

[30] D. M. Schnyer, P. C. Clasen, C. Gonzalez, and C. G. Beevers, "Evaluating the diagnostic utility of applying a machine learning algorithm to diffusion tensor MRI measures in individuals with major depressive disorder," *Psychiatry Res Neuroimaging*, vol. 264, pp. 1–9, Jun. 2017. https://doi.org/10.1016/j.pscychresns.2017.03.003

[31] N. Kannathal, U. R. Acharya, C. M. Lim, P. Sadasivan, and S. Krishnan, "Classification of cardiac patient states using artificial neural networks," *Exp Clin Cardiol*, vol. 8, no. 4, pp. 206–211, 2003.

[32] "Pathologists meet their match in tumour-spotting algorithm," *Nature*, vol. 561, no. 7724, pp. 436–437, Sept. 2018. https://doi.org/10.1038/d41586-018-06724-3

[33] L. Yao et al., "Application of artificial intelligence in renal disease," *Clinical eHealth*, vol. 4, pp. 54–61, Jan. 2021. https://doi.org/10.1016/j.ceh.2021.11.003

[34] J. Morley, M. Taddeo, and L. Floridi, "Google health and the NHS: Overcoming the trust deficit," *The Lancet Digital Health*, vol. 1, no. 8, p. e389, Dec. 2019. https://doi.org/10.1016/S2589-7500(19)30193-1

[35] S.-C. Huang, A. Pareek, S. Seyyedi, I. Banerjee, and M. P. Lungren, "Fusion of medical imaging and electronic health records using deep learning: a systematic review and implementation guidelines," *npj Digit. Med.*, vol. 3, no. 1, Art. no. 1, Oct. 2020. https://doi.org/10.1038/s41746-020-00341-z

[36] A. Rghioui, J. Lloret, M. Harane, and A. Oumnad, "A smart glucose monitoring system for diabetic patient," *Electronics*, vol. 9, no. 4, 4, Apr. 2020. https://doi.org/10.3390/electronics9040678

[37] C. K. M. Lee, M. N. Cheng, and C. K. Ng, "IoT-based asset management system for healthcare-related industries," *International Journal of Engineering Business Management*, vol. 7, pp. 7–19, Jan. 2015. https://doi.org/10.5772/61821

Chapter 9

Empowering autonomous vehicles in Industry 5.0 with digital twin technology

Krishna Vamshi Ganduri, Bhargav Prajwal Pathri, and Pinisetti Swami Sairam

9.1 INTRODUCTION

Emerging technologies such as Digital Twins (DT) and DT technologies have become increasingly important as the automotive sector transitions to Industry 5.0. It accentuates the human-centricity, sustainability, and seamless interaction between people and machines, Industry 5.0 asks for unique ideas that not only improve operational efficiency but also give value-driven operations priority [1, 2]. Among the most significant features of contemporary automobile engineering are autonomous cars capable of navigating and making judgments independently from human involvement as shown in Figure 9.1. Creating virtual clones of these vehicles thus rely on digital twin technology, which also enables real-time simulations, monitoring, and predictive analysis feasible. Combining information from many sources, including sensors and artificial intelligence systems, DT lets autonomous car systems continuously learn and improve. Driven in the era of Industry 5.0, this improves the performance, safety, and dependability of the vehicle, therefore impacting the direction of automotive engineering. Integration of DT technologies in autonomous cars will be essential to change mobility solutions globally as businesses strive toward more sustainable and smart transportation networks [3, 4, 5].

Digital Twin (DT) technology is a transformative innovation, creating virtual representations of physical systems and enabling real-time data integration for predictive maintenance and optimization. The digital twin is a concept that utilizes the interconnection of components, and the performance insights derived from process data. It is attracting increasing interest from firms and is becoming significant in the sector; however, it has not fully established itself. One cause is the absence of clarity on the construction of a digital twin. The subsequent section delineates the prerequisites and fundamental methodology for establishing a digital twin [6]. The fundamental framework, which is beneficial for generation, along with many distinct models, is examined. The necessary data for collection, as well as potential means for data transit and storage, are considered. Consequently, several platforms for creating a digital twin are analyzed and contrasted, and the

DOI: 10.1201/9781003568995-9

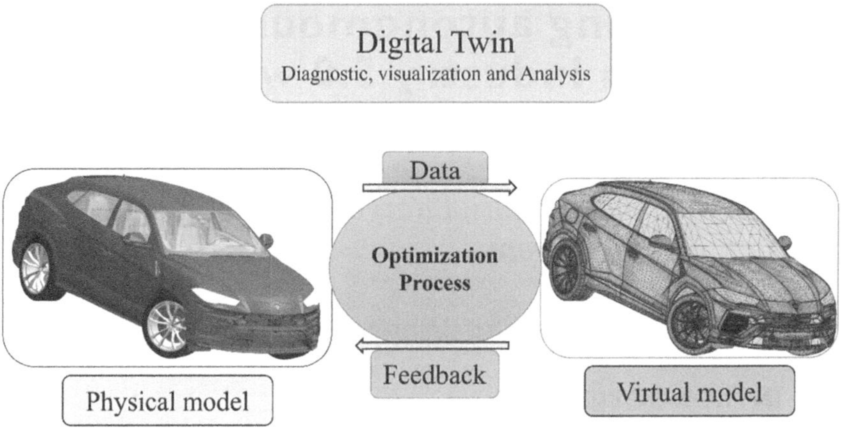

Figure 9.1 Digital twin model of autonomous vehicle.

distinct methodologies for developing a digital twin are elucidated. The benefits and drawbacks of various building methods are outlined based on their applications and business strategies [7].

9.1.1 History: Industry 1.0 to Industry 5.0

There have been five industrial revolutions to far, with the third and fifth being particularly essential for automation. While industry 5.0 will require human engagement with co-bots as they cooperate in the automation in the future, industry 4.0 is pushing cyber physical systems in the automation to increase vehicles efficiency shown in Figure 9.2.

Modern producers are under pressure to provide personalized products that cater to unique consumer wants, with lower batch sizes and fluctuating market demands [8]. They will require innovative technology and production methodologies that provide rapid adaptation to changes and the manufacturing of high-quality goods to meet these expectations. Computer Integrated Manufacturing (CIM) is a signal flow methodology connecting several integrated subsystems [9]. Over the past two decades, Enterprise Resource Planning (ERP) has emerged as the successor of Computer Integrated Manufacturing (CIM).

9.1.2 Digital twin

This section discusses the history of Digital Twin (DT) technology, including important concepts and dispelling some prevalent misconceptions about its identification. The notion of digital twins has been formalized since the early 2000s. However, shifting interpretations imply that the technology may

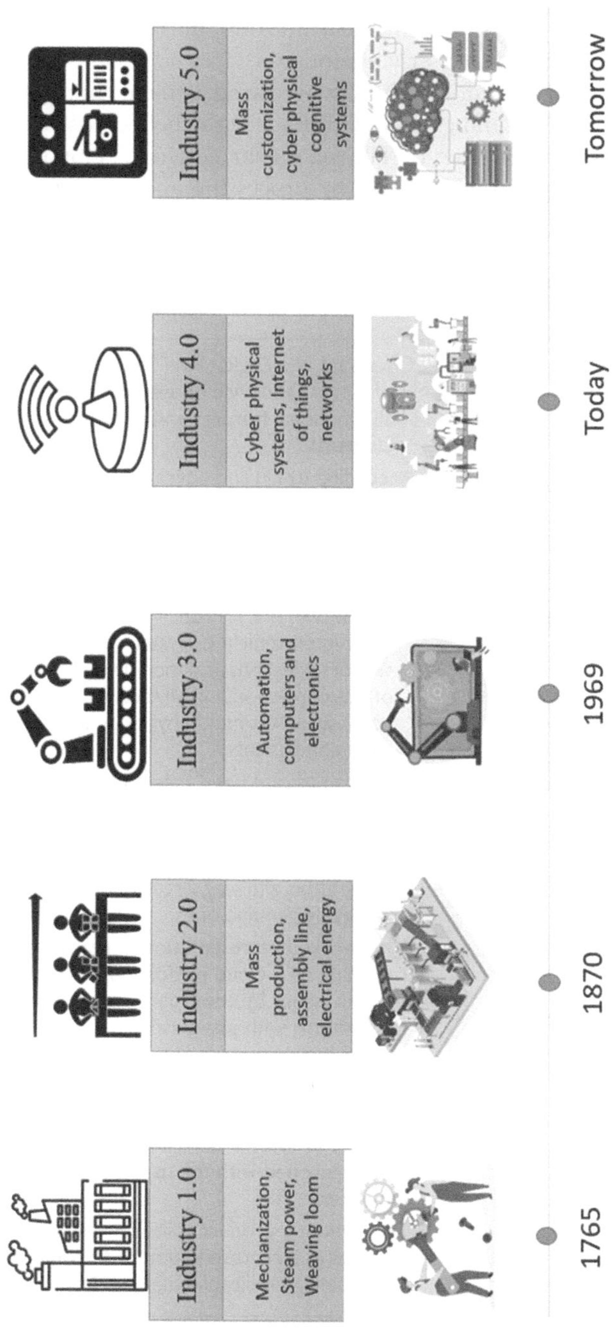

Figure 9.2 Industrial revolutions.

have been envisioned far earlier. Table 9.1 is an overview of Digital Twin definitions from several references. In essence, a Digital Twin is a real-time digital replica of a physical thing that not only mirrors but also optimizes the actual object using data-driven models.

Digital twin (DT) technology is a fast-growing, game-changing innovation that permits physical things to collaborate with their digital counterparts without issues. The Internet of Things, sensor networks, cloud computing, and big data analytics achieve this by processing and altering data in real

Table 9.1 Definitions of digital twins

Ref	*Author*	*Definition*
[10]	Chen	Chen said of a Digital Twin: "a computerized model of a physical device or system that represents all functional features and links with the working elements."
[11]	Liu et al.	According to Liu and colleagues, a Digital Twin is a "living model of the physical asset or system, which continually adapts to operational changes based on collected online data and can forecast the future of the corresponding physical counterpart."
[12]	Glaessgen and Stragel	A digital twin is a probabilistic simulation of a completed vehicle or system that uses data from sensors, ideal physical models, and the history of the corresponding physical item to mimic its lifespan. It is multi-physics, multi-scale, and all-encompassing.
[13]	Abdulmotaleb	The term "digital twin" refers to an electronic duplicate of any physical object, whether it be biological or not. Data may be easily transmitted between the physical and virtual worlds via integration, allowing virtual entities to live alongside genuine ones.
[14]	Lee et al.	A networked representation of real-world hardware running on a cloud platform that simulates health issues by combining data-driven analytical algorithms with pre-existing physical principles.
[15]	Soderberg et al.	Optimizing systems in real time utilizing digital models of actual systems.
[16]	Bolton et al.	The ever-changing digital model of a real-world object or system that can be understood, learned, and reasoned with the use of data collected in real-time.
[17]	Tao et al.	Throughout a product's lifespan, the digital twin serves as a representation of all components by combining physical, virtual, and interactive data.

time, thereby connecting the real and virtual worlds [18, 19]. The interface enables continuous data transmission throughout the lifetime of a product. A digital twin can learn about the changes to the operation and predict future performance [20]. DT systems synchronize physical assets with digital replicas through embedded sensors over advanced communication channels, providing performance, health, and maintenance histories. By projecting repair and service requirements using historical and real-time data, DT technology significantly aids in the transition from reactive to proactive strategies in predictive maintenance. In addition to this, it facilitates lifetime tracking through connected digital threads, enables monitoring and repair at a distance, and enhances operational optimization by bringing physical asset data together with IoT insights. Besides, digital twins can emulate aging and operation settings to assess asset lifespan, thus making the decision process more efficient. Digital twins constitute a robust platform integrating physical and virtual elements in sophisticated analytics to improve the performance of the system, optimize services, and ensure sustainable operations [21, 22].

The increasing use of digital technology has resulted in transformative digital twins, which are accurate digital replicas of physical systems or equipment. Several sectors and enterprises extensively use digital twin technology to identify issues and enhance productivity. Information technology breakthroughs, particularly the emergence of next generation ITs like the Internet of Things (IoT) [23], cloud computing, big data analytics, and artificial intelligence (AI), are driving the acceleration of digitalization. The convergence of the real and virtual worlds has made digitalization the primary catalyst for innovation across all sectors. The digitalization development shown in Figure 9.3 comprises four steps: digital enablement, digitalization assistance, digital control and linkage, and cyber-physical integration. The phrase "digital enablement" denotes the transformation of paper documents into digital formats while preserving their fundamental attributes [24, 25].

9.1.3 Role of IoT and AI in digital twin

The application of the IoT, DT, and AI modernizes autonomous vehicles and seamlessly integrates the virtual and physical worlds. To optimize performance for predictive maintenance and identify early defects, DT offers real-time monitoring and simulation of the vehicle's behavior. AI enhances decision-making by analyzing sensor data to forecast outcomes, while the Internet of Things (IoT) facilitates real-time communication between vehicles and their environment. The integration of autonomous vehicle systems with Industry 5.0 advances safety, reduces costs, optimizes energy consumption, and aids in achieving sustainability objectives [26, 27]. As a result,

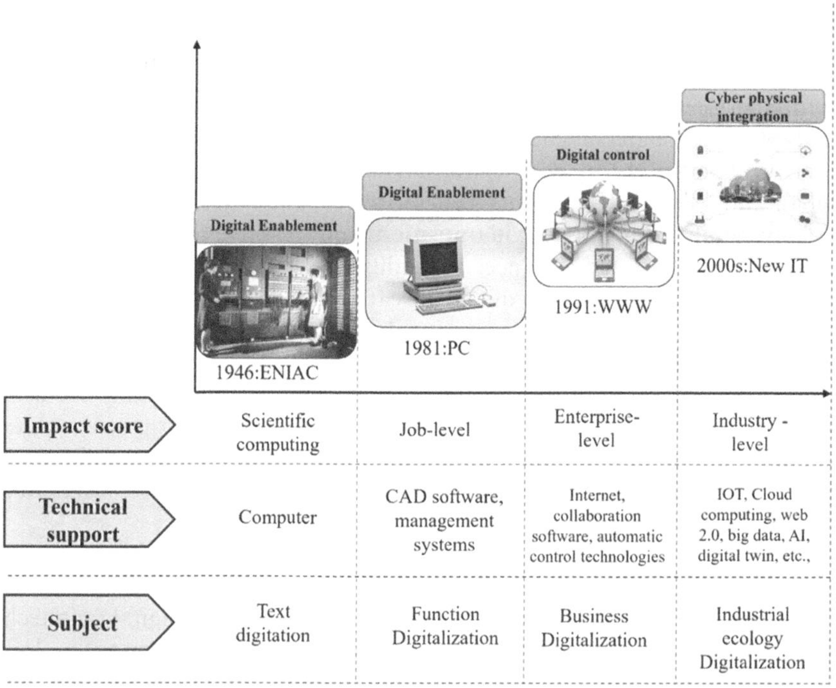

Figure 9.3 Evolution of digital paradigm.

digital twins drive significant improvements in performance and innovation within IoT and AI shown in Figure 9.4.

Digital twins are transforming manufacturing by generating virtual duplicates of actual assets, systems, and processes, allowing for real-time monitoring, simulation, and optimization of operations. These digital models improve operational efficiency by helping manufacturers to spot possible difficulties, streamline processes, and forecast equipment failures, resulting in less downtime and lower maintenance costs. Furthermore, digital twins enhance product quality by enabling virtual testing and design improvement, thereby reducing mistakes and waste. They help shorten time-to-market by reducing the need for physical prototypes and enabling quick changes throughout manufacturing. Furthermore, the data insights created by digital twins facilitate improved decision-making, allowing enterprises to adapt quickly to market needs and operational issues [28]. Digital twins, with their capacity to optimize resource utilization and encourage sustainable practices, are emerging as a vital instrument for improving smart manufacturing and driving the transition to Industry 4.

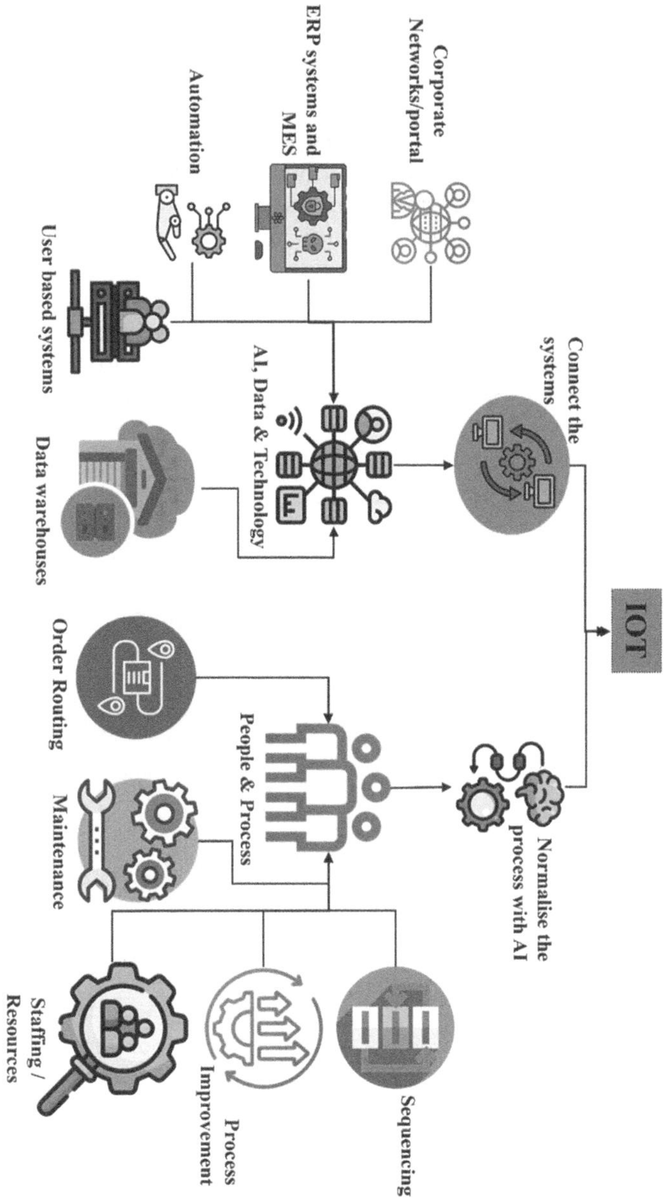

Figure 9.4 IoT, AI in digital twin.

9.1.4 Digital twin importance in Industry 4.0–Industry 5.0

Germany launched Industry 4.0 in 2011 with a focus on mass production and digitalization, while Industry 5.0 emphasizes sustainability, human-centricity, resilience, and human-robot collaboration as shown in Figure 9.5 [29, 30]. Industry 5.0 seeks applications in supply chain management, health, and smart cities, fully leveraging big data and semantic integration to maximize process efficiency. However, for achieving its vision, one must first see to the full potential of Industry 4.0. DT technology is critical in this shift. DT creates virtual representations of physical assets by integrating data from operational history, sensors, and physical models, thereby allowing real-time monitoring, simulation, and predictive maintenance. In production, DT assists in optimizing equipment performance, reduced risks, and the quality of final products. It has critical applications in smart city infrastructure, IoT security, and industrial processes. Recent research studies on DT indicate its applications in the Fifth Industrial Revolution (5IR) in terms of improving operational efficiency, troubleshooting, and testing

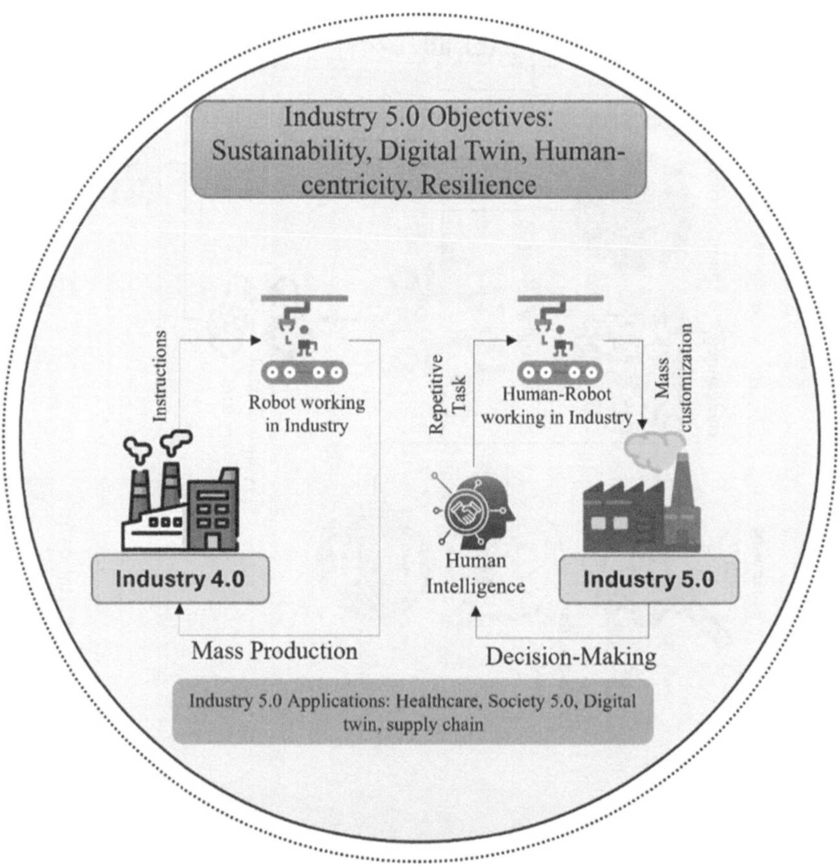

Figure 9.5 Concept of Industry 5.0.

within a virtual environment. A systematic review of 1,435 papers over a period of 2013–2022 yielded 136 key studies concerned with DT applications, architecture, and future directions in 5IR. The results form a technical roadmap for driving industrial innovation forward and effectively integrating DT technology into Industry 5.0 [31, 32].

9.1.5 Novelty

- Real-Time Predictive Intelligence: Continuous feedback and digital replicas enhance decision-making and predictive maintenance.
- Sustainability Focus: Optimizes energy use and minimizes emissions, in line with Industry 5.0's environmental objectives.
- IoT and AI Integration: A seamless link between IoT sensors and AI models allows for greater real-time data processing and autonomous driving.

9.2 RESEARCH QUESTIONS

The integration of Digital Twin (DT) technology in autonomous vehicles under the paradigm of Industry 5.0 presents a transformative opportunity to enhance efficiency, safety, and sustainability in transportation. However, the complexities of implementing this technology, coupled with its vast potential, necessitate a structured exploration of key challenges and opportunities. This section addresses pivotal research questions that underpin the role of DT in revolutionizing the automotive industry, focusing on the integration of IoT and AI as well as the challenges encountered in autonomous vehicle applications. Figure 9.6 illustrates the structured roadmap of the research, outlining the systematic approach taken to explore the role of DT technology in advancing Industry 5.0 and autonomous vehicles. The roadmap begins with a comprehensive literature review, highlighting the evolution of DT and its applications in the automotive industry.

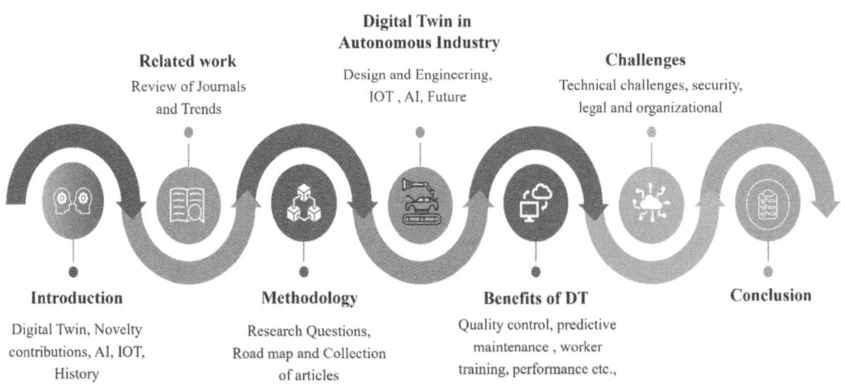

Figure 9.6 Roadmap of manuscript.

It transitions to the exploration of research questions, focusing on IoT and AI integration in autonomous systems. Key sections include the adopted methodology, challenges in implementation, and the transformative benefits of DT for manufacturers and end-users. The roadmap concludes with future directions, emphasizing sustainability, human-centric design, and scalability in Industry 5.0, ensuring a holistic understanding of DT technology's potential and limitations.

Complementing this, Figure 9.7 represents a comprehensive collection of research papers and studies analyzed to understand the current state, applications, and challenges of DT technology in various domains, with a particular focus on the automotive industry. This collection includes seminal works and recent advancements highlighting the evolution of DT, its integration with IoT and AI, and its transformative impact on manufacturing, supply chains, and autonomous vehicles. The figure visually categorizes these studies, emphasizing the diverse methodologies, findings, and areas of application. By synthesizing insights from these papers, the research identifies knowledge gaps and proposes innovative directions for future exploration in the realm of DT technology.

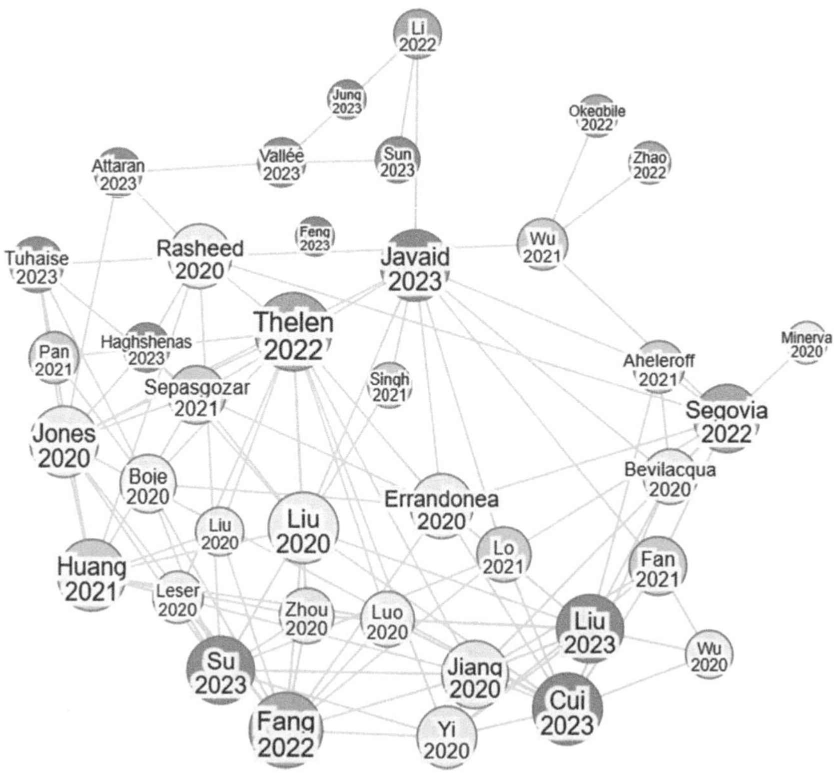

Figure 9.7 Collection of digital twin papers.

As the automotive industry embraces the transformative potential of Digital Twin (DT) technology within the Industry 5.0 paradigm, it becomes essential to address the key research questions that drive its development and application. Understanding how DT revolutionizes autonomous vehicles, its integration with IoT and AI, and the challenges that arise during implementation will provide a structured framework for advancing this field. These research questions not only explore the current capabilities of DT but also aim to chart a pathway for future innovations and practical applications in the automotive sector:

RQ1: In what ways is the Digital Twin revolutionizing the autonomous industry?
RQ2: How are IoT and AI integrating in autonomous vehicles?
RQ3: What are the challenges of Digital Twin in autonomous vehicles?

9.2.1 RQ1: In what ways is the Digital Twin revolutionizing the Autonomous Industry?

The automobile industry's market value has been increasing over the last several years, estimating an increase from $0.46 billion in 2020 to $5.06 billion in 2025. At the same time, it is expected that the market for digital twins would reach 73.5 billion US dollars by the year 2027. With the market's strong preference for digital twin technology as of the year 2024, let's have a look at the many uses of this technology that are now being used in the automobile sector [33].

Digital twins are becoming indispensable at many stages of car design and manufacturing, therefore changing the automobile industry. These virtual models help businesses to build and test prototypes before committing real production funds, therefore expediting the ideation process and facilitating early issue detection as shown in Figure 9.8. This approach greatly lowers expensive mistakes and speeds design deadlines. Digital twins may also be

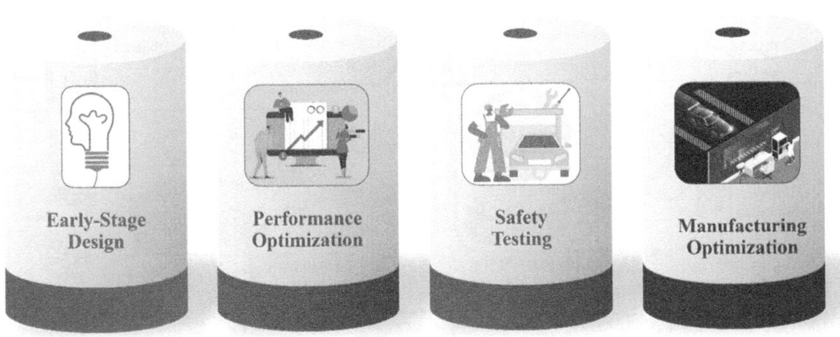

Figure 9.8 Digital twin in automotive design and engineering.

used by manufacturers to replicate vehicle performance in various environments, therefore enabling the optimal engine performance, aerodynamics, and fuel economy. Examining these features in a virtual environment helps companies decide what to do before actual manufacturing starts. The technology also promotes safety by means of simulated crash testing and safety evaluations, which ensure conformance to regulations and provide general vehicle safety features. Digital twins also streamline production processes by way of tooling designs, manufacturing techniques, and assembly line configurations, therefore reducing production costs and increasing productivity. Applied broadly, digital twin technology advances safety in the automotive industry's product development process, increases operational efficiency, and stimulates innovation [34].

9.2.2 RQ2: How IoT and AI integrating in autonomous vehicles?

In the automotive industry, a digital twin is the integration of an electronic equivalent of a real car into the manufacturing process. On the left, while driving in the real world, the physical automobile gathers data from all its numerous components, sensors, and surroundings. Access to this real-time data is critical for building and maintaining the digital twin, a mirror of the physical vehicle's real-time performance and condition, as shown in Figure 9.9 Real-time automotive data means that the computer Twin a computer replica of the car may alter in real time. This approach lets engineers duplicate the performance of the car, test it in numerous settings, and project issues even without operating it. Given constant information flow between the virtual and real cars, the digital twin might allow real-time monitoring and optimization. Data flow is Data arrows are important for this system. The digital twin tracks events, finds inefficiencies or abnormalities, and assesses performance using vehicle data. Though the story still develops, the digital twin created from using artificial intelligence to the data ends.

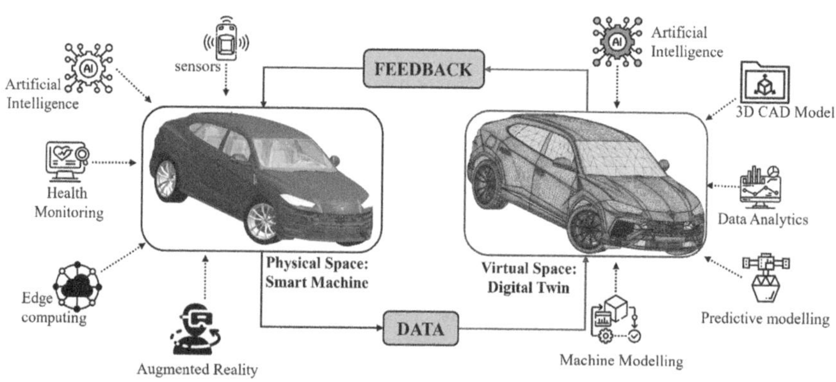

Figure 9.9 Integration of IOT and AI.

The feedback arrow illustrates the feedback loop, thereby stressing how the real car may learn concepts and ideas from its digital copy. The vehicle may increase its real-time efficiency by means of engine performance optimization, predictive maintenance, or operational process changes applying this feedback system. Artificial intelligence can predict vehicle faults by analyzing intricate data patterns, thereby enhancing driving safety and efficiency [35].

Advancement of autonomous driving systems rely mostly on the integration of artificial intelligence (AI) with Internet of Things (IoT) data so that they may make real-time decisions enhancing vehicle performance and safety. IoT devices incorporated in autonomous vehicles, such as sensors, cameras, and communication systems collect constantly massive amounts of data on the surroundings, traffic conditions, weather, and internal operations in this framework [36]. AI-powered analytics can manage this massive volume of IoT data in real-time especially using machine learning techniques. By use of sensor inputs, artificial intelligence can foresee possible hazards, optimize navigation courses, and change driving behavior based about affairs. For example, AI may advise the autonomous system to slow down or change lanes after seeing dangers like pedestrians or other automobiles instantly. IoT device data may similarly alert the system to unanticipated weather changes, therefore directing speed or vehicle handling for optimal safety. This perfect marriage of IoT data and artificial intelligence-driven decision-making also enables predictive analytics, where the AI system projects future events based on historical and real-time data. Predictive maintenance reduces the possibility of unexpected breakdowns by artificial intelligence forecasting when a part of the vehicle might fail using data from IoT sensors. Moreover, improving vehicle-to-vehicle (V2V) and vehicle-to-infrastructure (V2I) communication is artificial intelligence models learned on IoT data. These technologies may enhance traffic flow, reduce congestion, and prevent accidents by allowing autonomous automobiles to engage with smart city infrastructure in real time and with one other. IoT combined with artificial intelligence in autonomous driving systems is essentially a powerful combination that increases situational awareness, enables quick and precise decision-making, and produces a more efficient and safer driving experience.

9.2.3 Future of the digital twin in the automotive Industry

Digital twin technology offers a lot of potential for the future of automobile design and engineering, especially to emerging technologies like generative design, artificial intelligence (AI), and machine learning (ML). Among the areas digital twins are likely to change in the automobile industry are design, manufacturing, supply chain management, and maintenance services [37]. The automotive industry, traditionally reliant on physical prototypes characterized by protracted development timelines and cost-reduction efforts, is currently undergoing a transformation through the adoption of digital twins, which provide virtual validation and testing of various components.

Digital twins optimize process design and development by enhancing performance, safety, and efficiency through comprehensive simulation, thereby minimizing material waste and environmental impact. In the advancement of autonomous cars, digital twins enable artificial intelligence-driven simulations to assess and enhance self-driving systems efficiently and securely. Digital twins enhance supply chain efficiency through process modelling and optimization. This leads to decreased costs, waste, and lead times, while also allowing producers to swiftly adapt to market fluctuations [38].

9.3 BENEFITS OF DIGITAL TWIN IN AUTOMOTIVE MANUFACTURERS

- **Quality control enhancement**
  Digital twins allow real-time manufacturing process monitoring that allows businesses to quickly identify and resolve quality problems. BMW, for example, employs digital twins to find manufacturing bottlenecks, therefore increasing production, reducing downtime, and saving money while nevertheless preserving high standards of quality [18].
- **Predictive maintenance**
  Digital twins can predict maintenance needs before they occur by tracking product performance, therefore extending equipment lifetime and lowering unplanned downtime. Digital twins are used by General Motors (GM) to forecast maintenance needs, hence producing operating dependability and equipment longevity [39].
- **Worker training**
  Digital twins might be used by manufacturers to create realistic virtual simulations for worker training, therefore enhancing abilities and lowering the risk of mistakes. Ford uses digital twin technology in its VR training courses so staff members may develop their abilities in a controlled environment, hence enhancing operational precision and efficiency.
- **Performance monitoring**
  Digital twins let companies monitor production processes and automotive performance in real-time. Using digital twins, Tesla tracks attributes like battery life and energy consumption to deliver data to let customers better control vehicle performance and maintenance.
- **Supply chain management**
  Through real-time tracking of raw materials and final product transportation, digital twins help to enhance supply chain management. To streamline its supply chain and guarantee on-time part delivery, Toyota uses digital twins, for example, thereby improving inventory control and reducing delivery times.
- **Customization and personalization**
  Digital twin technology guarantees their practicality and affordability by allowing manufacturers to reproduce and assess choices on automobile customization including interior elements or paint colors.

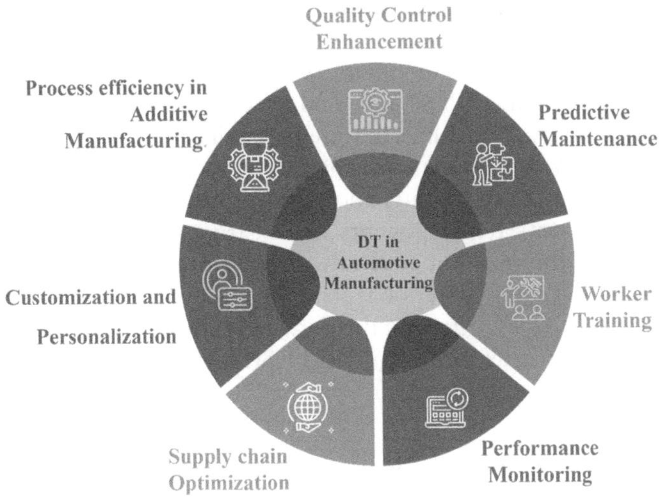

Figure 9.10 Benefits of DT in autonomous vehicles.

This flexibility provides tailored solutions depending on personal preferences, therefore improving client satisfaction and loyalty.

- **Process efficiency and manufacturing**
  In additive manufacturing, digital twins can mimic 3D printing techniques to uncover flaws and increase performance. By monitoring production in real time using IoT sensors, digital twins let companies find anomalies and improve process effectiveness. Figure 9.10 highlights the benefits of digital twin in autonomous vehicles.

9.3.1 RQ3: Challenges of digital twin in autonomous vehicles? (Figure 9.11)

Figure 9.11 Challenges of digital twins in the automotive industry.

9.4 TECHNICAL CHALLENGES

One of the key challenges of digital twin technology is keeping the data accurate and of high quality. Any data inaccuracy may lead to a poorly functioning twin as Model-based Systems Engineering (MBSE) and sensor fusion from many sources generate digital twins. This can damage both the car's performance and its maintenance procedures. For example, inadequate information on critical components like bearings or fuel injectors might hamper system optimization generally. Furthermore, compiling and merging data from different sources has its own set of challenges including privacy problems, data security, and the pure difficulty of processing such huge amounts of information [40]. Figure 9.11 presents the challenges of digital twin in automotive industry.

9.4.1 Security and privacy concerns

In digital twin systems particularly in automotive applications where sensitive data may contain trade secrets, personal customer information, and proprietary system details, data security and privacy are major issues. From harm of reputation to financial losses, illegal access represents a major concern. Businesses must apply strict security policies including access restrictions, firewalls, and encryption into use to prevent such threats. Moreover, especially in the manufacturing process using recently developed technologies like machine learning, following standards such as GDPR, CCPA, or HIPAA becomes necessary to secure consumer data and ensure legal conformity.

9.4.2 Organizational challenges

Using digital twins is further challenged by the opposition to change inside companies. Highly disruptive technologies called digital twins might drastically change processes, so some employees would be confused or hostile about how they would affect their responsibilities. Furthermore, using this technology calls for certain knowledge and expertise in fields like data analytics and software development, which could be difficult to come by. Cost issues also weigh as using digital twins calls for a significant hardware, software, and knowledge investment. For businesses ready to use digital twin technology, the possible return on investment including lower downtime and better equipment life—may make the initial outlay worthwhile [41].

9.4.3 Legal and regulatory issues

In the automotive industry, applying digital twin technology raises several legal and regulatory questions, particularly with respect to data protection and compliance. Digital twins largely rely on data from several sources, some of which might contain confidential business data, trade secrets, or consumer information—sensitive information like that. Following legislation like the General Data Protection Regulation (GDPR), California Consumer Privacy Act (CCPA), and Health Insurance Portability and Accountability Act (HIPAA), which become rather important, can help to protect data privacy and prevent legal penalties. Following these guidelines can be particularly difficult when new technologies such as machine learning are used in industrial processes as confirming compliance can get difficult [42].

9.4.4 Automotive industry digital twin challenges in India

Many challenges must be addressed if the Indian automobile industry is to fully exploit digital twin technology. Mostly contributing to these challenges are the nation's inadequate internet infrastructure and the use of antiquated technology. Many Indian companies still apply antiquated techniques that contradict digital twins and other contemporary data-driven solutions. Gathering and organizing the real-time data needed to create accurate digital models of vehicles or industrial processes might prove difficult given the weak data infrastructure now in place. Moreover, the interaction of digital twins with present systems generates various delays and inefficiencies resulting from interoperability problems. Time and money lost resulting from digital twins are already difficult for Indian automakers to manage. The lack of defined policies for data transfer and scattered data sources merely makes the problem much more difficult. To completely realize the possibilities of digital twin technology, the country will have to make significant expenditures in digital infrastructure, industry technical breakthroughs, and the acceptance of standard data protocols. We must overcome these challenges [43].

9.5 CONCLUSION

Digital twin (DT) technologies show great potential to revolutionize industry 5.0 autonomous cars. Combining IoT, artificial intelligence, and real-time sensor data lets DT construct a virtual mirror of the real automobile, therefore enhancing performance, safety, and sustainability. Early identification of any issues made possible by technology employs real-time simulations to help in boosting overall vehicle dependability and safety, therefore enabling early identification of any issues. Predictive maintenance is another crucial element as it ensures continual performance improvement while reducing running costs and downtime. Greatly lowering the need for actual prototypes, DT saves development time and expenses. Combining artificial intelligence with digital transformation within autonomous vehicles optimizes energy usage and supports the value-driven strategy of Industry 5.0 to help to accomplish sustainability targets. Through a customized and quick driving experience, it also enhances human-vehicle interactions. More broadly, by using logistical efficiency and material demand projections, DT strengthens automotive supply networks. Though it will reduce costs and drive innovation in automotive design, manufacturing, and operations, the combination of DT and autonomous vehicle technologies in Industry 5.0 will create a more efficient, sustainable, and safer future for transportation.

REFERENCES

[1] A. Verma et al., "Blockchain for Industry 5.0: Vision, Opportunities, Key Enablers, and Future Directions," *IEEE Access*, vol. 10, pp. 69160–69199, 2022, doi: 10.1109/ACCESS.2022.3186892

[2] M. Attaran and B. G. Celik, "Digital Twin: Benefits, Use Cases, Challenges, and Opportunities," *Decision Analytics Journal*, vol. 6, p. 100165, Mar. 2023, doi: 10.1016/j.dajour.2023.100165

[3] W. A. Ali, M. P. Fanti, M. Roccotelli, and L. Ranieri, "A Review of Digital Twin Technology for Electric and Autonomous Vehicles," *Applied Sciences*, vol. 13, no. 10, 2023, doi: 10.3390/app13105871

[4] S. Park, S. Maliphol, J. Woo, and L. Fan, "Digital Twins in Industry 4.0," *Electronics*, vol. 13, no. 12, pp. 1–16, 2024, doi: 10.3390/electronics13122258

[5] S. Ivanov, K. Nikolskaya, G. Radchenko, L. Sokolinsky, and M. Zymbler, "Digital Twin of City: Concept Overview," *Proc. - 2020 Glob. Smart Ind. Conf. GloSIC 2020*, no. 20, pp. 178–186, 2020, doi: 10.1109/GloSIC50886.2020.9267879S

[6] L. Wright and S. Davidson, "How to Tell the Difference Between a Model and a Digital Twin," *Advanced Modeling and Simulation in Engineering Sciences*, vol. 7, no. 1, 2020, doi: 10.1186/s40323-020-00147-4

[7] D. Adamenko, S. Kunnen, R. Pluhnau, A. Loibl, and A. Nagarajah, "Review and Comparison of the Methods of Designing the Digital Twin," *Procedia CIRP*, vol. 91, no. March, pp. 27–32, 2020, doi: 10.1016/j.procir.2020.02.146

[8] E. H. S. Robots, Universal, "Welcome to Industry 5.0," 2017.

[9] M. R. Pedersen et al., "Robot Skills for Manufacturing: From Concept to Industrial Deployment," *Robotics and Computer-Integrated Manufacturing*, vol. 37, pp. 282–291, 2016, doi: 10.1016/j.rcim.2015.04.002

[10] Y. Chen, "Integrated and Intelligent Manufacturing: Perspectives and Enablers," *Engineering*, vol. 3, no. 5, pp. 588–595, Oct. 2017, doi: 10.1016/J.ENG.2017.04.009

[11] Z. Liu, N. Meyendorf, and N. Mrad, "The Role of Data Fusion in Predictive Maintenance Using Digital Twin," 2018, p. 020023. doi: 10.1063/1.5031520

[12] E. H. Glaessgen and D. S. Stargel, "The Digital Twin Paradigm for Future NASA and U.S. Air Force Vehicles," *53rd AIAA/ASME/ASCE/AHS/ASC Struct. Struct. Dyn. Mater. Conf. 2012*, pp. 1–14, 2012.

[13] A. El Saddik, "Digital Twins: The Convergence of Multimedia Technologies," *IEEE Multimedia*, vol. 25, no. 2, pp. 87–92, Apr. 2018, doi: 10.1109/MMUL.2018.023121167

[14] J. Lee, B. Bagheri, and H.-A. Kao, "A Cyber-Physical Systems Architecture for Industry 4.0-Based Manufacturing Systems," *Manufacturing Letters*, vol. 3, pp. 18–23, Jan. 2015, doi: 10.1016/j.mfglet.2014.12.001

[15] R. Söderberg, K. Wärmefjord, J. S. Carlson, and L. Lindkvist, "Toward a Digital Twin for Real-Time Geometry Assurance in Individualized Production," *CIRP Annals*, vol. 66, no. 1, pp. 137–140, 2017, doi: 10.1016/j.cirp.2017.04.038

[16] R. N. Bolton et al., "Customer Experience Challenges: Bringing Together Digital, Physical and Social Realms," *Journal of Service Management*, vol. 29, no. 5, pp. 776–808, Nov. 2018, doi: 10.1108/JOSM-04-2018-0113

[17] F. Tao et al., "Digital Twin-Driven Product Design Framework," *International Journal of Production Research*, vol. 57, no. 12, pp. 3935–3953, Jun. 2019, doi: 10.1080/00207543.2018.1443229

[18] F. Biesinger, B. Kras, and M. Weyrich, "A Survey on the Necessity for a Digital Twin of Production in the Automotive Industry," *2019 23rd International Conference on Mechatronics Technology (ICMT)*, IEEE, Oct. 2019, pp. 1–8. doi: 10.1109/ICMECT.2019.8932144

[19] S. S. Shadrin, D. A. Makarova, A. M. Ivanov, and N. A. Maklakov, "Safety Assessment of Highly Automated Vehicles Using Digital Twin Technology," *2021 Intelligent Technologies and Electronic Devices in Vehicle and Road Transport Complex (TIRVED)*, IEEE, Nov. 2021, pp. 1–5. doi: 10.1109/TIRVED53476.2021.9639110

[20] Z. Wang et al., "Mobility Digital Twin: Concept, Architecture, Case Study, and Future Challenges," *IEEE Internet of Things Journal*, vol. 9, no. 18, pp. 17452–17467, Sep. 2022, doi: 10.1109/JIOT.2022.3156028

[21] Y. Zheng, S. Yang, and H. Cheng, "An Application Framework of Digital Twin and Its Case Study," *Journal of Ambient Intelligence and Humanized Computing*, vol. 10, no. 3, pp. 1141–1153, Mar. 2019, doi: 10.1007/s12652-018-0911-3

[22] J. Bao, D. Guo, J. Li, and J. Zhang, "The Modeling and Operations for the Digital Twin in the Context of Manufacturing," *Enterprise Information Systems*, vol. 13, no. 4, pp. 534–556, Apr. 2019, doi: 10.1080/17517575.2018.1526324

[23] J. Guo, M. Bilal, Y. Qiu, C. Qian, X. Xu, and K. K. Raymond Choo, "Survey on Digital Twins for Internet of Vehicles: Fundamentals, Challenges, and Opportunities," *Digital Communications and Networks*, vol. 10, no. 2, pp. 237–247, 2024, doi: 10.1016/j.dcan.2022.05.023

[24] S. Aheleroff, X. Xu, R. Y. Zhong, and Y. Lu, "Digital Twin as a Service (DTaaS) in Industry 4.0: An Architecture Reference Model," *Advanced Engineering Informatics*, vol. 47, p. 101225, Jan. 2021, doi: 10.1016/j.aei.2020.101225

[25] V. Kuts et al., "Digital Twin as Industrial Robots Manipulation Validation Tool," *Robotics*, vol. 11, no. 5, p. 113, Oct. 2022, doi: 10.3390/robotics11050113

[26] Q. Qi and F. Tao, "Digital Twin and Big Data Towards Smart Manufacturing and Industry 4.0: 360 Degree Comparison," *IEEE Access*, vol. 6, pp. 3585–3593, 2018, doi: 10.1109/ACCESS.2018.2793265

[27] G. Aceto, V. Persico, and A. Pescape, "A Survey on Information and Communication Technologies for Industry 4.0: State-of-the-Art, Taxonomies, Perspectives, and Challenges," *IEEE Communication Surveys and Tutorials*, vol. 21, no. 4, pp. 3467–3501, 2019, doi: 10.1109/COMST.2019.2938259

[28] L. Da Xu, E. L. Xu, and L. Li, "Industry 4.0: State of the Art and Future Trends," *International Journal of Production Research*, vol. 56, no. 8, pp.2941–2962, Apr. 2018, doi: 10.1080/00207543.2018.1444806

[29] M. Sharma, A. Tomar, and A. Hazra, "Edge Computing for Industry 5.0: Fundamental, Applications, and Research Challenges," *IEEE Internet of*

Things Journal, vol. 11, no. 11, pp. 19070–19093, 2024, doi: 10.1109/JIOT.2024.3359297

[30] A. Hazra, P. K. Donta, T. Amgoth, and S. Dustdar, "Cooperative Transmission Scheduling and Computation Offloading With Collaboration of Fog and Cloud for Industrial IoT Applications," *IEEE Internet of Things Journal*, vol. 10, no. 5, pp. 3944–3953, 2023, doi: 10.1109/JIOT.2022.3150070

[31] O. A. ElFar, C.-K. Chang, H. Y. Leong, A. P. Peter, K. W. Chew, and P. L. Show, "Prospects of Industry 5.0 in Algae: Customization of Production and New Advance Technology for Clean Bioenergy Generation," *Energy Conversion and Management: X*, vol. 10, p. 100048, Jun. 2021, doi: 10.1016/j.ecmx.2020.100048

[32] V. Damjanovic-Behrendt, "A Digital Twin-based Privacy Enhancement Mechanism for the Automotive Industry," in *2018 International Conference on Intelligent Systems (IS)*, IEEE, Sep. 2018, pp. 272–279. doi: 10.1109/IS.2018.8710526

[33] M. Jafari, A. Kavousi-Fard, T. Chen, and M. Karimi, "A Review on Digital Twin Technology in Smart Grid, Transportation System and Smart City: Challenges and Future," *IEEE Access*, vol. 11, pp. 17471–17484, 2023, doi: 10.1109/ACCESS.2023.3241588

[34] H. R. Chi, C. K. Wu, N.-F. Huang, K.-F. Tsang, and A. Radwan, "A Survey of Network Automation for Industrial Internet-of-Things Toward Industry 5.0," *IEEE Transactions on Industrial Informatics*, vol. 19, no. 2, pp. 2065–2077, Feb. 2023, doi: 10.1109/TII.2022.3215231

[35] M. Segovia and J. Garcia-Alfaro, "Design, Modeling and Implementation of Digital Twins," *Sensors*, vol. 22, no. 14, p. 5396, Jul. 2022, doi: 10.3390/s22145396

[36] M. Roccotelli, M. Nolich, M. P. Fanti, and W. Ukovich, "Internet of Things and Virtual Sensors for Electromobility," *Internet Technology Letters*, vol. 1, no. 3, May 2018, doi: 10.1002/itl2.39

[37] P. Augustine, *The Industry Use Cases for the Digital Twin Idea*, 1st ed., Elsevier Inc., 2020. doi: 10.1016/bs.adcom.2019.10.008

[38] G. Bhatti, H. Mohan, and R. Raja Singh, "Towards the Future of Smart Electric Vehicles: Digital Twin Technology," *Renewable and Sustainable Energy Reviews*, vol. 141, p. 110801, May 2021, doi: 10.1016/j.rser.2021.110801

[39] A. F. Mendi, "A Digital Twin Case Study on Automotive Production Line," *Sensors*, vol. 22, no. 18, p. 6963, Sep. 2022, doi: 10.3390/s22186963

[40] A. Hazra, "Promising Role of Visual IoT: Challenges and Future Research Directions," *IEEE Engineering Management Review*, vol. 51, no. 4, pp. 169–178, 2023, doi: 10.1109/EMR.2023.3304121

[41] A. Rasheed, O. San, and T. Kvamsdal, "Digital Twin: Values, Challenges and Enablers From a Modeling Perspective," *IEEE Access*, vol. 8, pp. 21980–22012, 2020, doi: 10.1109/ACCESS.2020.2970143

[42] T. Y. Melesse, V. Di Pasquale, and S. Riemma, "Digital Twin Models in Industrial Operations: A Systematic Literature Review," *Procedia Manufacturing*, vol. 42, no. 2019, pp. 267–272, 2020, doi: 10.1016/j.promfg.2020.02.084

[43] C. Kober, M. Fette, and J. P. Wulfsberg, "Challenges of Digital Twin Application in Manufacturing," *2022 IEEE International Conference on Industrial Engineering and Engineering Management (IEEM)*, IEEE, Dec. 2022, pp. 0162–0168. doi: 10.1109/IEEM55944.2022.9989654

Chapter 10

Augmented reality reshaping medical and engineering education

Manvi Mishra, Brijmohan, and Harsh Vardhan Saxena

10.1 INTRODUCTION

The term augmented reality is used to describe the process of blending computer graphics with real situations. AR interfaces enable users to view a combination of real-world and virtual images and real-world locations and objects simultaneously through a portable or head-mounted display—which can be transparent or with a video feed, it's about the environment—unlike fully immersive VR It can overload the image-processing AR interfaces that explore surroundings through an augmented reality interface surrounding them that enhances the real-world experience [1], unlike other computer interfaces that tend to draw images users away from reality and onto screens.

AR has a big impact on how we teach engineering and medicine. It shows students hard ideas in a way they can touch and play with, using 3D models they can see and move around. This helps them grasp and recall tricky stuff better [2]. For engineers in training, AR lets them test out their designs and fix problems in a fake world, which keeps them out of harm's way as they learn [3]. In the same way medical students can practice operations and figure out what's wrong with patients without putting real people at risk. AR also tells students right away if they're doing things right so they pick up skills faster. It connects book learning to real life by showing how ideas work. What's more, AR can change to fit how each person learns best, so students can go at their own speed. It even makes it possible to learn from far away so more people can get an education. With AR, both engineering and medical classes can pull students in more, get them more involved, and teach them better. That's why we need AR in these areas.

Single-user AR interfaces are particularly made for one person and have been evolved for a lot of programs, such as scientific visualization, production, and computer-aided practice. These uses highlight how AR interfaces can allow users to engage with the physical environment in new and previously impossible approaches. A latest technology called Mixed Reality (MR) combines aspects of virtual and augmented fact to supply an easy and engaging enjoy, that unifies the real and digital worlds [4].

DOI: 10.1201/9781003568995-10

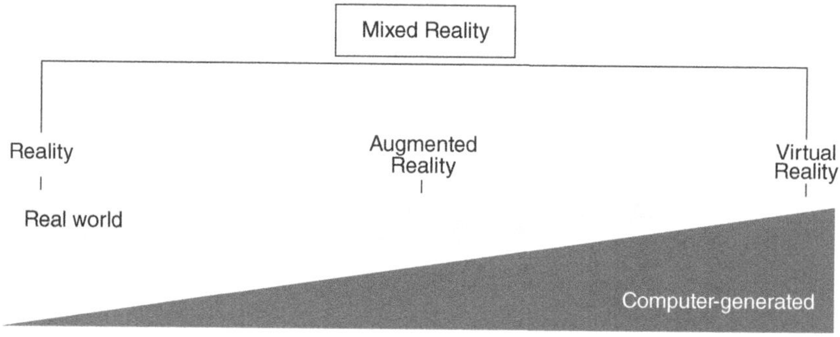

Figure 10.1 Continuum of reality and virtuality.

More immersive and cohesive experiences exists in MR settings because virtual items interact with and react to the real world in real time, in addition to being superimposed onto it as in AR [5]. The Figure 10.1 shows the continuum of reality and virtuality depicting the finest description of MR, which is the superset of Augmented and Virtual Reality. The real environment is the well-known physical reality in which we all live and in which our perceptions correspond with our experiences in the physical world [6].

AR improves our experiences in the real world by overlaying digital data in the physical environment [7]. In contrast, virtual reality (VR) puts people in a state of computer-generated existence, fully simulated. VR users are disconnected from the outside world and only engage with the virtual world [8]. Example VR 360 Tours etc.

Medical and engineering education face several roadblocks that make learning tough. Students often can't get hands-on practice because of safety worries or not enough resources, which is a big problem. It's hard for students to grasp complex ideas when they're shown in two dimension or three dimension. Also, students don't get quick feedback on how they're doing, which makes it hard for them to improve continuously.

The persistent gap between theoretical understanding and practical application is what hampers the students' capacity to convert ideas into workable solutions. Another challenge is meeting the students' varied learning preferences and demands. Moreover, geographical barriers limit access to education for some students. AR solves these problems. AR enhances the efficiency and accessibility of medical and engineering education through the provision of simulated environments for practical experience, allowing interaction with 3D models to enhance understanding, giving instant feedback, filling the gap between theory and action, accommodating individual learning styles, and allowing remote learning.

Traditional educational methods often struggle to engage students fully, especially when dealing with complex subjects like human anatomy or engineering principles. Our motivation is to create a learning tool that captivates and immerses users, making education an interactive and enjoyable experience.

10.2 LITERATURE REVIEW

There are many research papers on AR-based systems available [9]. Table 10.1 lists a few of them. The literature review discusses how AR is transforming education and how advanced technologies can enhance learning [10]. Key research papers highlight various applications and implications of AR are:

Caudell, T.P. and Mizell, D.W. (1992) [11], Bajura, M., et al (1992) [12], Feiner, S., et al. (1993) [13] and Azuma, R.T.(1997) [14] showcase work based on AR in medical and education.

Mark Billinghurst's (2002) [15] paper discusses AR's transformative potential across educational and collaborative settings, showcasing its ability to overlay virtual imagery onto real-world environments to enrich learning and streamline collaborative tasks. Zhou F et al. (2008) [16] Analyze market scenario for AR over last ten years. Dunleavy, Dede, & Mitchell's (2009) [17] highlighted the promise of AR in education but also raised issues such as mentor/ teacher preparation, pedagogical integration, and content creation.

Phil Diegmann and Manuel Schmidt-Kraepelin (2015) [18] discusses the various advantages of AR usages in education domain and offer frameworks for understanding its impact across different contexts in education and learning areas.

Zhu et al. (2016) [19] assesses a critical analysis on the use of AR in anatomy education and discusses the possible benefits for the improvement of students' understanding of the structure of the human body and obstacles in practical implementation.

Singh et al. (2019) [20] discuss the use of augmented reality in education. Hanid et al. (2020) [21] explore new learning mechanism with AR in education domain. Latest evaluations by Prit Kaur and Alok Mantri (2022) [22], Murat Akçayır and Gökçe Akçayır (2023) [23], Dorota Kamińska (2023) [24] and K. C. Nakul and Praveen Kumar (2024) [25] Discuss developments, trends, opportunities, and enduring obstacles to acceptance and efficacy as you continue to evaluate AR's educational implications.

In general, the literature shows how AR is an evolving tool to be used in the development of engagement, better learning outcomes, and diverse educational needs, with the ongoing difficulties in cost effective implementation and 24/7 accessibility.

The current education approach, which emphasizes the integration of Augmented Reality AR with computer-aided instruction in a variety of subjects including medical, biology, anatomy and teaching of engineering courses, has garnered great interest. However, there exists a considerable knowledge gap: the cost of AR devices and their deployment in educational environments are often unconsidered. This cost barrier hinders the accessibility of AR-enhanced education resources and discourages thorough integration.

Table 10.1 Summary of analyzed papers based on AR

S. No.	*Year*	*Paper title*	*Reference*	*Domain*	*Findings*
1	1992	AR in Manual Manufacturing	Caudell, T.P. and Mizell, D.W.	Augmented Reality	Draws attention to how head-up display technology can be used in manual production processes.
2	1992	Merging Virtual Objects with the Real World	Bajura, M., et al.	Augmented Reality	Results shows techniques for combining ultrasound images inside a patient with the real world.
3	1993	Knowledge-Based AR	Feiner, S., et al.	Augmented Reality	Illustrates the creation of a real-world knowledge-based graphic annotation system.
4	1997	A Survey of Augmented Reality	Azuma, R.T.	Augmented Reality	An overview and survey of developments in augmented reality technology.
5	2002	AR in Education	Mark Billinghurst	Augmented Reality	Depicts challenges of AR Hardware
6	2008	Trends in AR: A Review of ISMAR	Zhou, F., et al.	Recent Trends in Augmented Reality	An analysis of market trends for AR over the past ten years
7	2009	AR Applications in Education	Dunleavy, M., et al.	AR in Education	An overview of the research on the usage of AR in the classroom.
8	2015	Benefits of AR in Educational Environment	Diegmann, Phil, and Schmidt-Kraepelin, Manuel	AR in Education	Highlights the advantages of AR in learning scenario

9	2016	AR for Anatomy Education	Zhu, E. et al.	Immersive medical educational experiences with AR	An extensive analysis of AR's use cases in anatomy instruction.
10	2019	The Use of Augmented Reality in Education: A Comprehensive Review	Gaganpreet Singh and Ajay Mantri	Interactive digital overlays in teaching	Reviews the existing research on AR application in education sector.
11	2020	Learning strategies using AR Technology in Education: Meta-Analysis	Hanid et al.	AR in learning	Explores learning methods with AR technology in education
12	2022	Enhancing Learning Experiences with Augmented Reality: A Review of Educational Implications	Prit Kaur and AlokMantri	Digital augmentation in the classroom	Analyze ways to improve learning situations that have educational relevance.
13	2023	Augmented Reality in Education: A Systematic Review of Literature	Murat Akçayır and GökçeAkçayır	AR in Education	To offer a thorough analysis of the developments and trends of augmented reality in education.
14	2023	Augmented Reality in Education: Trends, Perspectives, and Opportunities	DorotaKamińska et al.	Technology-enhanced education through AR	Provides guidance for researchers and opportunities.
15	2024	Virtual and AR Applications	Nakul K. C, Praveen Kumar	AR & VR in Engineering Education	The study examines the growing conversation around AR and VR, especially in technology and general studies, emphasizing their many uses.

Our study aims to address this issue by offering an innovative solution that allows students to use AR-enhanced content alongside normal learning materials, focusing on anatomy, engineering, and medicine. Moreover, our approach is cost-effective, ensuring that both teachers and students can afford the resources. Such research serves to narrow the divide between the theoretical advancements of AR education and its effective use, thereby optimizing educational outcomes.

In our work Feature Extraction Using Scale Invariant Feature Transform (SIFT) Algorithm is used as elaborated in Section 10.2.1

10.2.1 Feature extraction using SIFT algorithm

Existing AR systems employ feature extraction algorithms which are highly sensitive to factors such as scale and orientation i.e: images are similar in nature. Because we have shown through various applications that it is more effective than anything else, feature extraction is being done by this study with SIFT. The robust feature points are extracted and are invariant to projective transformation, restricted affine distortion, image rotation, scaling, and changes in lighting. The following are the basic steps involved in the feature extraction with SIFT mechanism [26]:

The feature points are detected using a SIFT detector by looking for local maxima using the difference of Gaussian (DoG) at various image scales. For example, let $f(x,y)$ be the input, and the appropriate scale space may be described as follows:

$$\begin{aligned} L(x,y,\sigma) &= G(x,y,\sigma) * f(x,y) \\ G(x,y,\sigma) &= \left(1/2\pi\sigma^2\right)e^{-\left(x^2+y^2\right)/2\sigma^2} \end{aligned} \tag{10.1}$$

$$\begin{aligned} D(x,y,\sigma) &= \left(G(x,y,k\sigma) - G(x,y,\sigma)\right) * f(x,y) \\ &= L(x,y,k\sigma) - L(x,y,\sigma) \end{aligned} \tag{10.2}$$

where σ indicates the scale, is the variable-scale Gaussian function, and * is the convolution operator. The difference of two adjacent scales, separated by a multiplicative factor k, is used to calculate the DoG for stable feature points as follows:

As seen in Figure 10.2, a Gaussian pyramid is used to depict the procedure. Every sample point is compared to all of its neighbors in order to identify the local maxima and minima of $D(x,y,\sigma)$. The central point in Figure 10.3, for instance, is contrasted with its nine neighbors at the upper and lower neighboring scales, and its 8 neighbors at the same scale. The candidate feature points are the estimated extreme values.

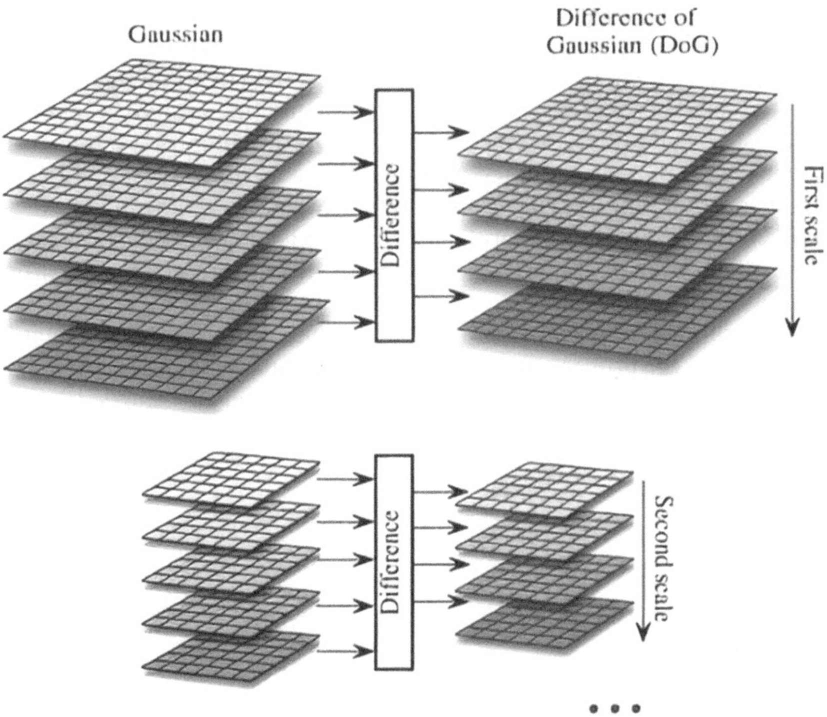

Figure 10.2 Creating a scale-space pyramid with features that undergo scale invariance.

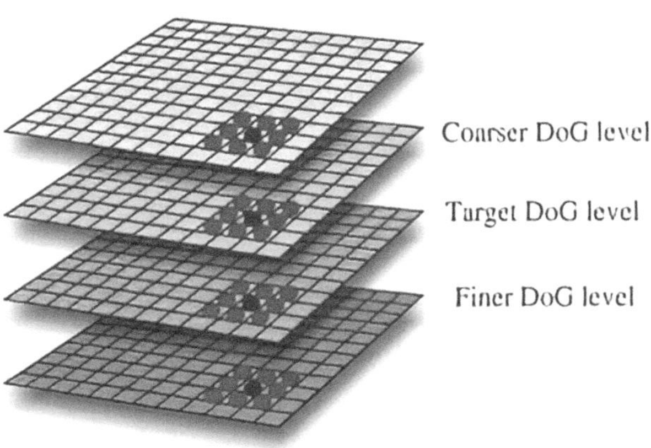

Figure 10.3 Comparing for extreme value points based on features that are scale invariant.

The final points depend only on the correspondence of the possible points; Accurate principal areas are obtained by a Taylor series expansion of the scale area; If the strength within this limit is less than a certain threshold, the main component is removed because the system does not have a minimum contrast; In addition, edge effects are removed to improve stability and enhance anti-noise capabilities; The sampling is performed around the key point location, and the direction of the histogram is used to measure the gradient direction of neighboring pixels; Subsequent changes in orientation, size, and location of a transformed image result in invariants in the transformation process [27].

A neighbor is chosen around the key point, then a local coordinate is formed with the primary direction of key point at Zero degrees to maintain rotational invariance [28]. An instance is presented in Figures 10.4 and 10.5 [28].

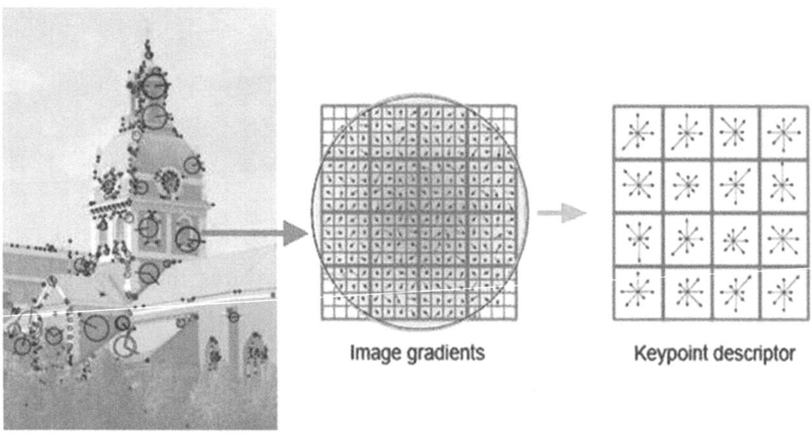

Figure 10.4 Feature point extraction [27].

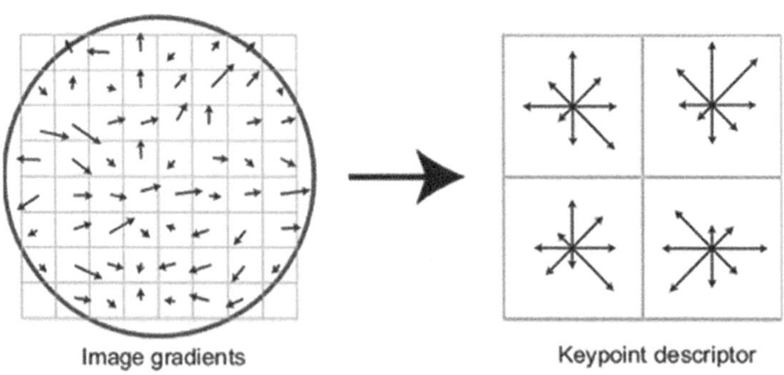

Figure 10.5 Generation procedure of SIFT feature descriptor [28].

10.3 PROPOSED WORK

The primary objective of this proposed studies is to expand a today's AR utility the usage of Vuforia and the Unity engine that would efficiently depict the anatomical structure of a human for medical students and crucial engineering diagrams for technocrat. Suggested app goal is to provide customers with an interesting and interactive experience that permits them to explore and recognize the intricacies of the human body's anatomical components and rigorous engineering ideas.

The software design facilities on marker-based AR, wherein users utilize their tool's camera to test a delegated marker. Upon a success marker detection, a three-dimensional version representing anatomical systems is seamlessly augmented into the real-global environment. This dynamic integration permits customers to interact with virtual anatomical fashions overlaid onto physical surfaces, thereby facilitating a unique and captivating learning enjoy. The workflow of proposed AR application is shown in Figure 10.6.

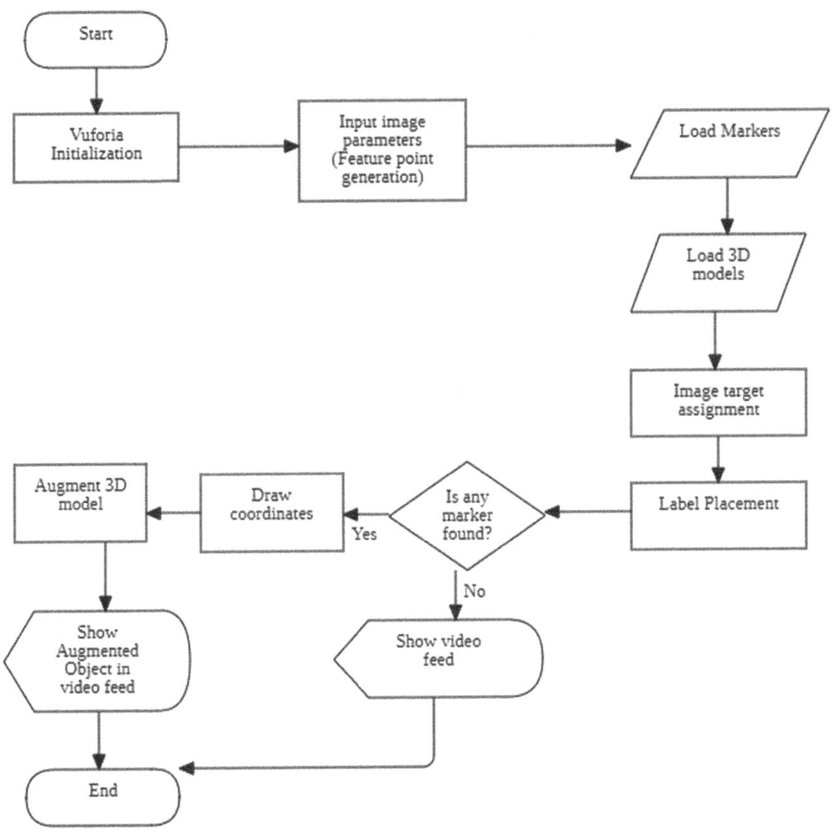

Figure 10.6 Workflow of proposed AR medical application.

The suggested AR app allows end users to engage with augmented three dimensional model. Information about the roles and importance of each component is available to users by tapping on different labels associated with anatomical structures. The interactive labeling system is an educational tool that will allow users to explore the organs' functional characteristics and obtain a thorough understanding of their responsibilities in the body of a person.

The very first step towards making an AR anatomy learning application is Vuforia setup, which includes creating an account on the developer dashboard so that those essential tools become available. For generation of markers for AR tracking, relevant anatomical images are uploaded to the dashboard, keeping in mind feature points and tracking suitability. A marker database is designed by downloading the information from Vuforia, which then lays down the foundation for marker-based AR tracking.

Moving on to Unity integration, a project is created, and Vuforia Engine is integrated into the development environment. The AR camera in Unity is configured with Vuforia so that it will calibrate correctly and align itself with the physical world. Image targets are then implemented, which involves importing the created marker database into Unity as picture targets. Markers are connected to comparable anatomical structures in a hierarchy.

Next is the integration of 3D models, either by creating or importing anatomical structures. These models are attached as children to image targets to ensure proper alignment and scaling. When creating a system for informative labeling on 3D models, user interactions are taken into account to ensure a smooth learning process.

A user interface (UI) for transformation control is then designed and implemented. This UI allows users to control the transformation of augmented 3D models, including scaling, rotation, and translation, enhancing the interactive learning experience. Finally, the AR anatomy learning app is prepared for deployment on relevant educational platforms, ensuring compatibility with commonly used devices.

While browsing the AR Content, Students can explore a diverse range of augmented reality content, including labeled 3D models and interactive simulations as depicted in Figure 10.7. Students have the capability to engage with realistic simulations, facilitating a hands-on learning experience in anatomical structures and engineering components. Students can monitor their learning progress within the application, gaining insights into completed modules or achievements. Students can access and interact (i.e. Rotate/Scale Up-Down) with labeled 3D models. Educators have the same capability as students to explore augmented reality content, including labeled 3D models and interactive simulations. The same task will be done by Admin or educator also. Beside this he can upload the education resources such as supplementary materials or specific content tailored to their curriculum.

Existing AR systems employ feature extraction algorithms which are highly sensitive to factors such as scale and orientation. SIFT technique is

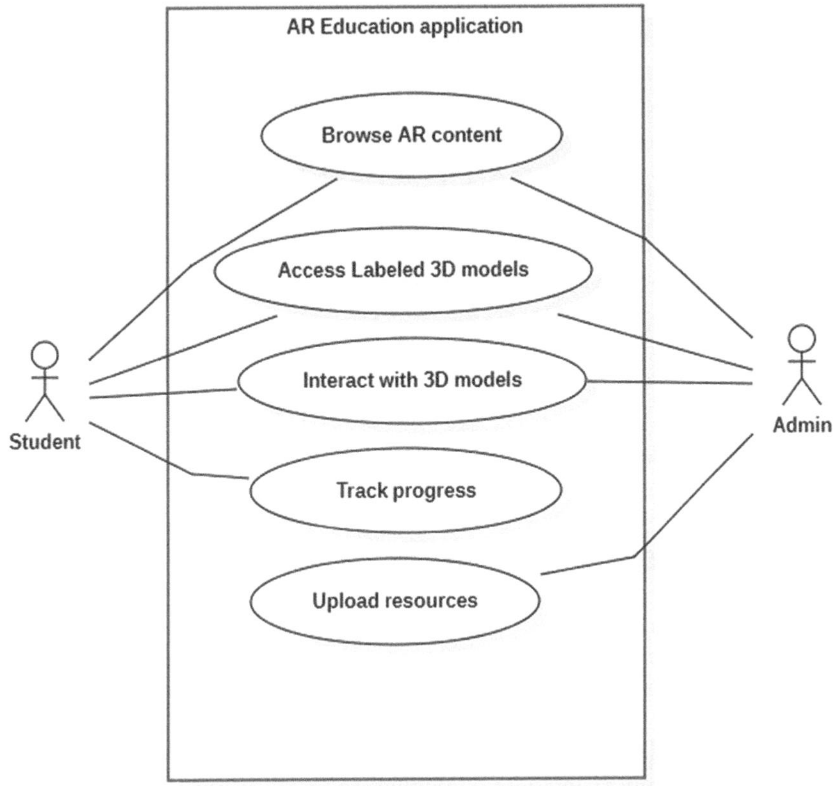

Figure 10.7 Use case diagram of proposed AR Medical application.

employed for feature extraction due to its demonstrated superiority over alternative methods in various computer vision applications. SIFT makes use of Difference of Gaussians (DoGs) which have necessary scaling built in, hence features are extracted from the markers which remain the same no matter how much worse the quality of the image gets, this ensures enhanced tracking [28].

Firstly, Preprocessing is done while moving towards feature extraction. Then Key points of images are identified with Difference of Gaussians (DoG) algorithm. It highlights regions in the image where there are significant intensity changes. These regions become the key points. Once key points are identified, orientation is assigned to each of them. This step is crucial for ensuring invariance to image rotation. It helps maintain consistency in recognizing the same key point under different orientation. Eventually descriptors are computed for each key point based on the gradient information around the key point. Gradient provide a way to capture the local structure and characteristics of the image near the key point. Box 10.1 represents pseudo code of SIFT algorithm used in the work.

Box 10.1 Pseudocode of SIFT algorithm

```
function SIFT(image):
   // Step 1: Preprocess the image
preprocessed Image = preprocess Image(image)
   // Step 2: Detect key points using the Difference of
   Gaussians (DoG)
key points = detect Key points(preprocessed Image)
   // Step 3: Assign an orientation to each key point
key points = assign Orientation(key points, preprocessed
Image)
   // Step 4: Compute descriptors based on gradient
   information
   descriptors = compute Descriptors(preprocessed Image,
   key points)
   return key points, descriptors
```

10.4 RESULTS AND DISCUSSION

For implementing the research work windows 11 OS is utilized with x86, x64 architecture along with SSE2 instruction set. Graphic API OpenGL 4.3 and DX12 is used. Some open source softwares are also used shown in Table 10.2.

Some of the development functions that the simplified code editor Visual Studio Code support include debugging, task execution, and version control. Blender, a free and open-source 3D creation tool, which supports modeling, rigging, animation, simulation, rendering, compositing, motion tracking, and even video editing and game development. Writing code that communicates with your objects, importing and assembling components, making or importing animations and so many things are also possible in Unity.

Feature points contribute to real-time updates in AR experiences. As the device moves, the system continuously tracks and updates the positions of

Table 10.2 Open-source software for AR based applications

Technology	*Description*
Unity	Game development engine
Visual studio code	Development environment
Vuforia engine	AR engine
Blender	3D modeling software

feature points, ensuring that the augmented content remains synchronized with the physical environment. Figure 10.8 represented the feature point extraction process. Table 10.3 highlighted the Metadata and Augment ability rating of the marker while working on AR application.

In augmented reality (AR), labeling and marker image target assignment are essential steps to enhance the real-world environment with digital information or objects. Labeling involves associating digital content or information with real-world objects or scenes. It helps AR systems understand and recognize specific objects, enabling them to overlay relevant digital information on the physical world. Marker-based AR relies on predefined images or patterns (markers) that act as triggers for AR content. Assigning digital content to marker images allows the AR system to recognize these markers and superimpose virtual elements on them. To do so mapping is done between detected markers and corresponding digital content. Pseudocode for image based tracking is given in Box 10.2.

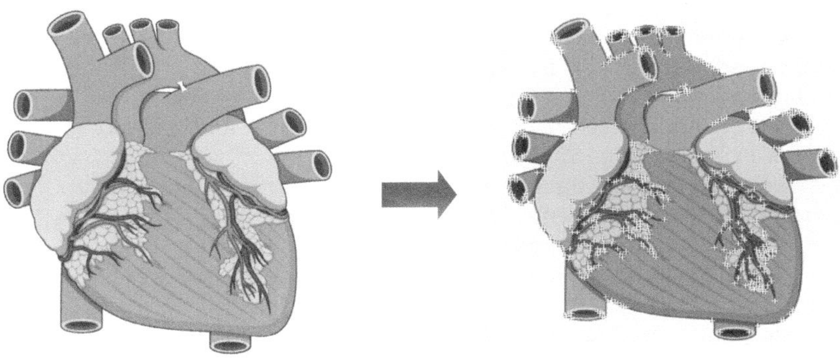

Figure 10.8 Feature Point extraction.

Table 10.3 Metadata and augment ability rating of the marker

Type	*Image*
Status	Active
Target ID	18d9b1e000b24a9783262ab3a93600c
Augmentable	5/5 stars
Added	Nov 16, 2023 11:04
Modified	Nov 16, 2023 11:04

Box 10.2 Pseudocode for image based tracking

```
import arcore # Import the ARCore library
# Define marker images
marker_images = {
   "marker1": "path/to/marker1.jpg",
   "marker2": "path/to/marker2.jpg",
   # Add more markers as needed
}
# Create AR session
ar_session = arcore.create_session()
# Load marker images
for marker_id, marker_path in marker_images.items():
ar_session.load_marker(marker_id, marker_path)
# Start AR session
ar_session.start()
# Main AR loop
while ar_session.is_running():
   # Capture camera feed
camera_frame = ar_session.get_camera_frame()

   # Detect markers in the camera feed
detected_markers = ar_session.detect_markers
(camera_frame)
   # For each detected marker, associate and render
   corresponding digital content
   for marker in detected_markers:
marker_id = marker.get_id()
digital_content = load_digital_content(marker_id)
ar_session.render_content(digital_content, marker.
get_transform())
# Stop AR session when done
ar_session.stop()
```

The result of this phase is depicted in Figure 10.9. Final desired heart application is shown in Figure 10.10.

The human body 3D model in AR app is shown in Figure 10.11. For Engineering students, radial pneumatic engine is also designed by AR based approach as shown in Figure 10.12.

The culmination of work is a trio of meticulously labeled 3D models: the heart, full body, and radial engine. Crafted with precision and detail, these models serve as comprehensive educational tools for research work.

The radial pneumatic engine is a type of power plant that operates using compressed air. Its 3D model typically features multiple cylinders arranged in a circular pattern around a central crankshaft. Compressed air is injected into each cylinder in sequence, causing pistons to move and generate rotational motion. This configuration provides high power efficiency and is commonly used in applications requiring compact but robust propulsion systems.

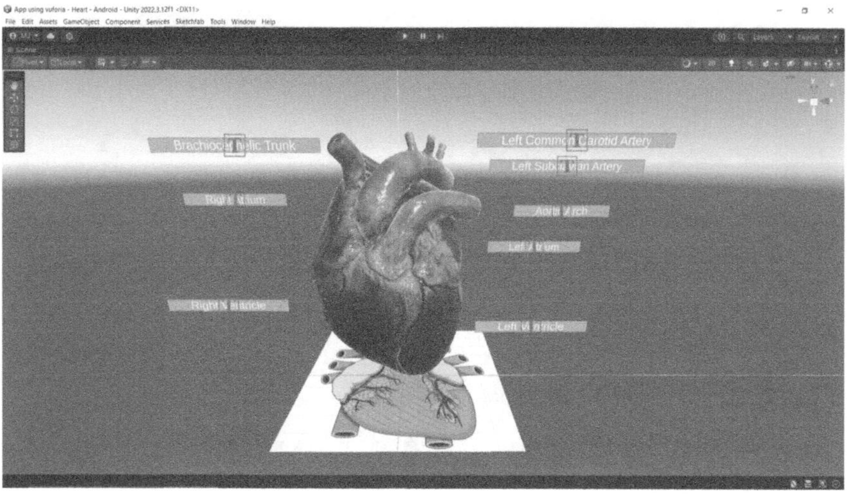

Figure 10.9 Labeled 3d model of Heart and Marker Image target assignment.

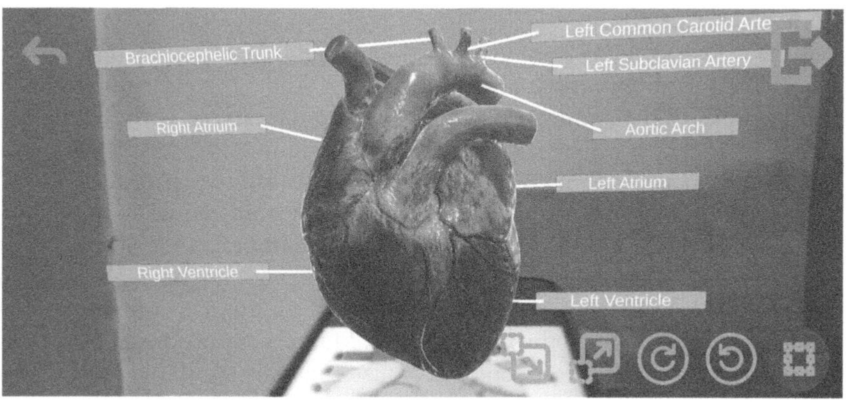

Figure 10.10 Final application labeled placement.

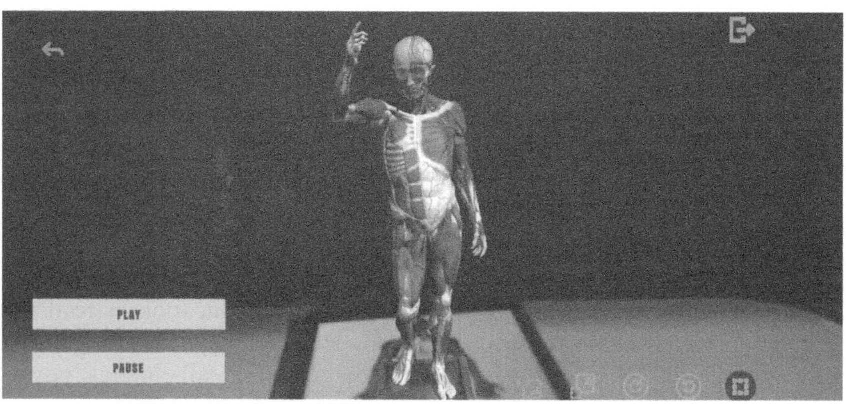

Figure 10.11 Human body 3d model in AR application.

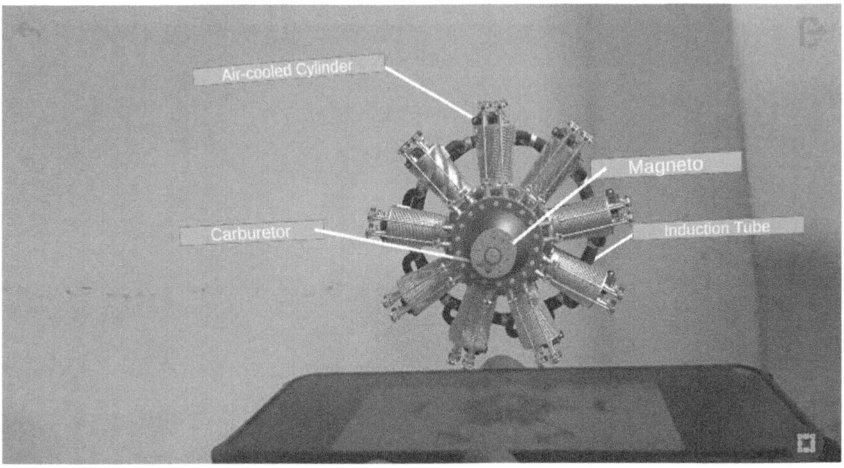

Figure 10.12 Radial engine 3d model in AR application.

10.5 CONCLUSION

Augmented Reality in medical teaching allows students to learn better with specific links in the application, which permits interaction with 3D models of human body parts. The paper aims to present how AR can enhance learning about human anatomy. To do so, Vuforia and a Unity platform is utilized which fuels an AR application. This application would enable an individual to view 3D models of body components in anatomy medical courses as well as engine assemblies in the field of engineering and learn about them through an interactive way. Based on Vuforia, an app can impose virtual models on the real world through markers. Markers are produced by special points that the application could recognize thus putting the models in the correct location.

The Augmented Reality application objective is to maximize user interaction by intelligent embedding hyperlinks in its end user interface, thereby allowing users to provide an interactive and smooth experience exploring anatomical models. In this respect, this application resorted to a very clean and organized layout to enhance a lot of navigation and interaction through an intuitive learning environment. A separate module to train the marker is included in the application; thus, structured learning as part of the curriculum is allowed. The application further extends its functionality with the addition of a marker-less model recognition feature, which increases the extent of engaging learning experiences. This approach is aimed at not only capturing the attention of users but also aligning the application with educational norms, providing a versatile tool for anatomy education in academic settings. With AR based medical application students may learn tough topic of anatomy or other complex engineering diagrams in really interesting and very efficient way, hence, the application helps out faculties, mentors as well as students by learning crucial structures in highly easy manner.

REFERENCES

[1] Liarokapis, F. (2004). *Web3D and AR to support engineering education.* The World Institute for Engineering and Technology Education (WIETE), pp 11–14.

[2] Dhar, P., Rocks, T., Samarasinghe, R. M., Stephenson, G., & Smith, C. (2021). Augmented reality in medical education: Students' experiences and learning outcomes. *Advances in Health Sciences Education*, 26(8), pp 1–9.

[3] Martin, S. et al. (2011). New technology trends in education: seven years of forecasts and convergence. *Computers & Education*, 57(3), 78–85.

[4] Kyaw, B. M., Saxena, N., Posadzki, P., Vseteckova, J., Nikolaou, C. K., & George, P. P. (2019). Augmented reality and mixed reality for healthcare education beyond surgery: An integrative review. *MDPI*, 11, 1–6.

[5] Bronack, S.C. (2011). The role of immersive media in online education. *Journal of Continuing Higher Education*, 59(2), 45–52.

[6] Arvanitis, T.N. et al. (2007). Human factors and qualitative pedagogical evaluation of a mobile augmented reality system for science education used by learners with physical disabilities. *Personal and Ubiquitous Computing*, 13(3), 123–130.

[7] Hsiao, K. H., & Chen, R. F. (2019). An exploration of augmented reality-based contextual learning: A case study of medical education. *Educational Technology & Society*, 22(3), 239–252.

[8] Wu, H. K., Lee, S. W. Y., Chang, H. Y., & Liang, J. C. (2013). Current status, opportunities and challenges of augmented reality in education. *Computers & Education*, 62, 41–49.

[9] Persky, A. M., & Pollack, G. M. (2017). Transforming experience, accelerating outcomes: An integrative review of the role of technology in medical education. *Teaching and Learning in Medicine*, 29(3), 314–323.

[10] Albrecht, U. V. (2018). Opportunities and challenges of augmented reality in education: A systematic review of the literature. *Journal of Educational Technology & Society*, 21(2), 222–236.

[11] Caudell, T.P., & Mizell, D.W. (1992). Augmented reality: An application of heads-up display technology to manual manufacturing processes. In *Proceedings of the Twenty-Fifth Hawaii International Conference on Systems Science*, Kauai, Hawaii, 7th–10th Jan. 1992, Vol. 2, pp. 659–666.

[12] Bajura, M., Fuchs, H., Ohbuchi, R. (1992). Merging virtual objects with the real world: Seeing ultrasound imagery within the patient. In Proceedings of SIGGRAPH '92. ACM Press, pp. 203–210.

[13] Feiner, S., MacIntyre, B., & Seligmann, D. (1993). Knowledge-based augmented reality. *Communications of the ACM*, 36(7), 53–62.

[14] Azuma, R. T. (1997, August). A survey of augmented reality. *Presence: Teleperators and Virtual Environments*, 6(4), 355–385.

[15] Billinghurst, M. (2002). *Augumented reality in education.* New Horizon for Learning.

[16] Zhou, F., Duh, H.-L., & Billinghurst, M. (2008). Trends in augmented reality tracking, interaction and display: A review of ten years in ISMAR. Mixed and Augmented Reality, *ISMAR 7th IEE/ACM International Symposium*, pp. 193–202. IEEE.
[17] Dunleavy, M., Dede, C., & Mitchell, R. (2009). Augmented reality applications in education: A literature review. *Journal of Science Education and Technology*, 40(2), 7–21.
[18] Diegmann, P., and Schmidt-Kraepelin, M. (2015). Benefits of AR in educational environment. *AISel*, 14–16.
[19] Zhu, E., Hadadgar, A., Masiello, I., Zary, N., &Amini, M. (2016). Augmented reality for anatomy education: A systematic review. *Anatomical Sciences Education*, 12, 565–582.
[20] Singh, G., & Mantri, A. (2019). The use of augmented reality in education: A comprehensive review, 6, 17–19.
[21] Hanid, M. F. A., Said, M. N. H. M., & Yahaya, N. (2020). Learning strategies using augmented reality technology in education: Meta-analysis. *Universal Journal of Educational Research*, 8(5A), 51–56.
[22] Kaur, D. P., Mantri, A., & Horan, B. (2020). Enhancing student motivation with use of augmented reality for interactive learning in engineering education. *Procedia Computer Science*, 172, 881–885.
[23] Prit Kaur, D., Mantri, A., & Horan, B. (2022). Design implications for adaptive augmented reality based interactive learning environment for improved concept comprehension in engineering paradigms. *Interactive Learning Environments*, 30(4), 589–607.
[24] Kamińska, D., et al. (2023). Augmented reality: Current and new trends in education. *Electronics*, 12(16), 3531.
[25] Nakul, K. C., & Kumar, P. (2024). Virtual and augmented reality applications: A broader perspective. *Journal of Data Science and Intelligent Systems*, 10–13.
[26] Gonzalez, R. C., Woods, R. E., & Eddins, S. L. (2004). *Digital image processing using MATLAB*. Pearson Education India.
[27] Bag-of-features descriptor on SIFT Features with OpenCV (BoF-SIFT). https://www.codeproject.com/Articles/619039/Bag-of-Features-Descriptor-on-SIFT-Features-with-O
[28] Lowe, D. G. (2004, Jan.). Distinctive image features from scale-invariant key points. *International Journal of Computer Vision*, 60(2), 1–25.

Index

Pages in **bold** refer to tables.